T0341172

Advanced R Solutions

Series Editors

Recently Published Titles

Learn R: As a Language
Pedro J. Aphalo

Using R for Modelling and Quantitative Methods in Fisheries
Malcolm Haddon

R For Political Data Science: A Practical Guide
Francisco Urdinez and Andres Cruz

R Markdown Cookbook
Yihui Xie, Christophe Dervieux, and Emily Riederer

Learning Microeconometrics with R
Christopher P. Adams

R for Conservation and Development Projects: A Primer for Practitioners
Nathan Whitmore

Using R for Bayesian Spatial and Spatio-Temporal Health Modeling
Andrew B. Lawson

Engineering Production-Grade Shiny Apps
Colin Fay, Sébastien Rochette, Vincent Guyader, and Cervan Girard

Javascript for R
John Coene

Advanced R Solutions
Malte Grosser, Henning Bumann, and Hadley Wickham

For more information about this series, please visit: https://www.crcpress.com/Chapman--HallCRC-The-R-Series/book-series/
CRCTHERSER

Advanced R Solutions

Malte Grosser
Henning Bumann
Hadley Wickham

CRC Press is an imprint of the
Taylor & Francis Group, an **informa** business
A CHAPMAN & HALL BOOK

First edition published 2022
by CRC Press
6000 Broken Sound Parkway NW, Suite 300, Boca Raton, FL 33487-2742

and by CRC Press
2 Park Square, Milton Park, Abingdon, Oxon, OX14 4RN

CRC Press is an imprint of Taylor & Francis Group, LLC

ISBN: 978-1-032-00750-2 (hbk)
ISBN: 978-1-032-00749-6 (pbk)
ISBN: 978-1-003-17541-4 (ebk)

DOI: 10.1201/9781003175414

Publisher's note: This book has been prepared from camera-ready copy provided by the authors.

Authors' note: Because there were no exercises in chapters 1, 12, 16, 17 and 22 in *Advanced R, Second Edition* (ISBN 9780815384571), these chapters are excluded from this book.

Malte

To Elena

Henning

To my family

Contents

Preface

Welcome to *Advanced R Solutions*!

This book provides worked-out solutions to the exercises given in *Advanced R* [Wickham, 2019c] and reflects our efforts to understand and acquire its content.

Advanced R covers R [R Core Team, 2020] and programming. It presents the specific characteristics of the R language to programmers and helps R users to improve their understanding of general programming concepts.

When we came across *Advanced R*, it gave us a more thorough understanding of the R code we worked with daily and helped us to see the underlying principles more clearly. The content helped us to get a more complete picture of R's programming landscape.

We soon re-discovered that reading about programming is not enough and that it helps greatly to open the editor and write some code along the way. The clear structure of *Advanced R* and the exercises given provided a great starting point for this.

We think of this book as a solutions manual, which intends to supplement and support your own study of the R language through *Advanced R*. We hope that it will help you to stay on track and allow you to check your understanding and progress along the way. The solutions may also give you another perspective on some of the presented material.

How this book came to be

The solutions in this book are written from our personal perspective and current level of understanding. We both came from mathematics and statistics backgrounds preparing us more carefully for data analysis than for programming. So, we were R users first and improved as programmers in our day jobs and through working on projects like this one.

By taking advantage of the `{bookdown}` package [Xie, 2016] to structure our process, we created this book almost as a by-product. While the initial progress was fun and motivating, actually providing solutions to all of the 284 exercises took slightly longer than originally expected (and Hadley's rewrite of *Advanced R* halfway in between didn't really make the journey shorter).

As the project matured, we strived to provide solutions that were as clean, straightforward, and expressive as possible. As well-written code is often more expressive than lengthy explanations, many of the solutions are rather code heavy. The written explanations aim to provide context and motivation, discuss important implementation details, or relate to the practical work of being an R programmer.

Hadley Wickham wrote *Advanced R* and created the exercises which form the substructure of this book. We took the task to solve them as correctly and idiomatically as possible. When we finished a chapter, we asked Hadley to review it. His feedback included many comments (which we then had to resolve), corrections, and suggestions, as well as a few complete solutions. We repeated this process until each exercise was reviewed and approved. As a result, we feel pretty good about the quality of the solutions in the book. However, any remaining mistakes or inconsistencies are certainly on us.

How to use this book

Since this book builds so heavily on *Advanced R*, we think it should be read together with the textbook, either as a hardcopy or the online version (`https://adv-r.hadley.nz`). Working on an exercise first by yourself should in general give you the biggest benefit.

It may be a good idea to start with the parts of *Advanced R* that are most relevant to your work and interest. You can certainly read the book cover to cover, but we think that you don't have to, though it's probably a good idea to start with the foundations part.

Of the more difficult exercises in the book, only a few were completed in one go. Often we had to reread the question or look up the related content in *Advanced R* and started by writing a few lines of code first or consulting the documentation. Reading the source code (preferably with syntax highlighting) and searching the web were typically quite helpful.

To support your study, you may also be interested in the R4DS Advanced R book club (`https://GitHub.com/r4ds/bookclub-Advanced_R`), where groups of readers regularly discuss a different chapter of *Advanced R*.

In case you want to do more or have a special interest in the mechanics of base R, you may be interested in checking out the first edition of *Advanced R* (`http://adv-r.had.co.nz/`) [Wickham, 2014]. Some additional solutions related to that edition can be found at `https://advanced-r-solutions-ed1.netlify.app/`.

There is one recommendation from *Advanced R* that we'd like to echo: reading source code can be tremendously helpful in developing your programming skill! For example, you can just head to GitHub and start looking into the source code of packages you love and developers you admire. While reading, it's not necessary to understand every line of code right away. Keeping this a regular practice (for a while) will expose you to many new ideas, patterns, and design choices, and also expand your R vocabulary.

We don't necessarily apply many of the concepts taught in *Advanced R* in daily programming and that's okay! But we hope that the code we write has become more robust, expressive, and readable, and it's actually quite easy to see the progress when we take a look at the earlier drafts of our own code.

Acknowledgements

Many open source projects are the result of the work of a lot of people; so is this. We would like to explicitly mention and thank everybody who contributed solutions, raised questions, or helped to fix spelling and grammar to improve this work:

@3zhang, Anne (@ahoffrichter), Anh N Tran (@anhtr), Arash (@arashHaratian), Leon Kim (@BetweenTwoTests), Jun Cai (@caijun), @Charles926, Safouane Chergui (@chsafouane), Corrado Lanera (@CorradoLanera), @davidblitz, Zhuoer Dong (@dongzhuoer), @Elucidase, Fabian Scheipl (@fabian-s), @HannesOberreiter, @its-gazza, Jorge Aranda (@jorgearanda), @lotgon, @MajoroMask, Maya Gans (@MayaGans), Øystein Sørensen (@osorensen), Peter Hurford (@peterhurford), @philyoun, PJ (@pieterjanvc), Robert Krzyzanowski (@robertzk), Emily Robinson (@robinsones), Tanner Stauss (@tmstauss), @trannhatanh89, and Yihui Xie (@yihui).

Tobias Stalder (`@toeb18` (`https://twitter.com/toeb18`)) designed the beautiful cover, which visualizes the structure of *Advanced R* and its exercises.

Thanks to CRC Press for the interest in the project and our editor, Rob Calver, and his assistant, Vaishali Singh, for their patience and support in making this book a reality.

Thanks to our managers and companies for granting us some flexibility with our work schedules and generally supporting the completion of this project.

Conventions

A brief overview of conventions we followed and decisions we made.

- Some chapters and sections in *Advanced R* do not contain exercises. In our book you will see that we skipped these chapters and sections. This decision introduces some gaps in the numbering, but we feel that keeping the numbers in sync with those of *Advanced R* will provide the most practical value.
- We strived to follow mostly the tidyverse style guide (`https://style.tidyverse.org/`). The `{styler}` package [Müller and Walthert, 2020] helped us to check many of the rules automatically.
- Each chapter of this book was rendered in a separate R session via the `{bookdown}` package. We configured this process to initially:
  - set `` `%>%` <- magrittr::`%>%` `` to unlock the pipe operator without specifically loading the `{magrittr}` package [Bache and Wickham, 2020] every time,
  - set a random seed (`1014`) to improve reproducibility (similar as in *Advanced R*), and
  - define a few `{ggplot2}` and `{knitr}` options.

  You can check out the exact code (`https://GitHub.com/Tazinho/Advanced-R-Solutions/blob/main/common.R`) on GitHub.
- We chose to keep the code in this book as self-contained as possible.
  - The packages used are usually loaded in the beginning of each chapter.
  - We repeat all code from *Advanced R* that is necessary to work on an exercise but not explicitly part of the exercise. When some longer code passages (from *Advanced R*) are omitted, this is explicitly stated in the solution.
- The printed version of the book was rendered with R version 4.0.3 (2020-10-10) and the most recent available package versions as of December 2020. (The print version of *Advanced R* was rendered with R version 3.5.2.)
- Emoji images in the printed book come from the open-licensed Twitter Emoji (`https://github.com/twitter/twemoji`).
- Benchmarks are computed when the book is rendered. While this improves reproducibility, the exact results will depend on the system creating the document.

Closing remarks

We are so happy to finish this exciting project that in fact neither of us really had the time for. We probably wouldn't have made it to the finish line if we hadn't worked on it together.

Collaboration is powerful and it's fun to build and share. The various backgrounds represented in the R community generally make this exchange much more interesting and meaningful. Much of this success is possible because R is free software. At least in theory, everyone can contribute and no one can take away your freedom to do so.

The automated systems we build using these tools are not neutral and the rapid adoption of data-driven processes in business and technology does clearly affect our everyday lives and societies. It's important that everyone has a fair say in the discussions about these systems and participates in their design. Against this background, we chose to donate half of our royalties from this book to `https://rladies.org/`, an organization empowering minority genders in the R community.

Thank you for your interest in this project and we hope the solutions will be of value to you.

See you around!

Malte Grosser `@malte_grosser` (`https://twitter.com/malte_grosser`)

Henning Bumann `@henningsway` (`https://twitter.com/henningsway`)

Part I

Foundations

2

Names and values

Prerequisites

In this chapter we will use the {lobstr} package [Wickham, 2019a] to help answer questions regarding the internal representation of R objects.

```
library(lobstr)
```

2.2 Binding basics

Q1: Explain the relationship between `a`, `b`, `c`, and `d` in the following code:

```
a <- 1:10
b <- a
c <- b
d <- 1:10
```

A: `a`, `b`, and `c` point to the same object (with the same address in memory). This object has the value `1:10`. `d` points to a different object with the same value.

```
list_of_names <- list(a, b, c, d)
obj_addrs(list_of_names)
#> [1] "0x5639a82a2d70" "0x5639a82a2d70" "0x5639a82a2d70"
#> [4] "0x5639a838f808"
```

Q2: The following code accesses the mean function in multiple ways. Do they all point to the same underlying function object? Verify this with `lobstr::obj_addr()`.

DOI: 10.1201/9781003175414-2

```
mean
base::mean
get("mean")
evalq(mean)
match.fun("mean")
```

A: Yes, they point to the same object. We confirm this by inspecting the address of the underlying function object.

```
mean_functions <- list(
  mean,
  base::mean,
  get("mean"),
  evalq(mean),
  match.fun("mean")
)

unique(obj_addrs(mean_functions))
#> [1] "0x5639a43823e0"
```

Q3: By default, base R data import functions, like `read.csv()`, will automatically convert non-syntactic names to syntactic ones. Why might this be problematic? What option allows you to suppress this behaviour?

A: Column names are often data, and the underlying `make.names()` transformation is non-invertible, so the default behaviour corrupts data. To avoid this, set `check.names = FALSE`.

Q4: What rules does `make.names()` use to convert non-syntactic names into syntactic ones?

A: A valid name must start with a letter or a dot (not followed by a number) and may further contain numbers and underscores ("_"s are allowed since R version 1.9.0).

Three main mechanisms ensure syntactically valid names (see `?make.names`):

1. Names that do not start with a letter or a dot will be prepended with an "X".

   ```
   make.names("")    # prepending "x"
   #> [1] "X"
   ```

   The same holds for names that begin with a dot followed by a number.

```r
make.names(".1")  # prepending "X"
#> [1] "X.1"
```

2. Additionally, non-valid characters are replaced by a dot.

```r
make.names("non-valid")  # "." replacement
#> [1] "non.valid"
make.names("@")          # prepending "X" + "." replacement
#> [1] "X."
make.names("  R")        # prepending "X" + ".." replacement
#> [1] "X..R"
```

3. Reserved R keywords (see `?reserved`) are suffixed by a dot.

```r
make.names("if")  # "." suffix
#> [1] "if."
```

Interestingly, some of these transformations are influenced by the current locale. From `?make.names`:

> The definition of a letter depends on the current locale, but only ASCII digits are considered to be digits.

Q5: I slightly simplified the rules that govern syntactic names. Why is `.123e1` not a syntactic name? Read `?make.names` for the full details.

A: `.123e1` is not a syntactic name, because it starts with one dot which is followed by a number. This makes it a double, `1.23`.

2.3 Copy-on-modify

Q1: Why is `tracemem(1:10)` not useful?

A: When `1:10` is called an object with an address in memory is created, but it is not bound to a name. Therefore, the object cannot be called or manipulated from R. As no copies will be made, it is not useful to track the object for copying.

```
obj_addr(1:10)  # the object exists, but has no name
#> [1] "0x5639a786c700"
```

Q2: Explain why `tracemem()` shows two copies when you run this code. Hint: carefully look at the difference between this code and the code shown earlier in the section.

```
x <- c(1L, 2L, 3L)
tracemem(x)

x[[3]] <- 4
```

A: Initially the vector `x` has integer type. The replacement call assigns a double to the third element of `x`, which triggers copy-on-modify.

```
x <- c(1L, 2L, 3L)
tracemem(x)
#> <0x66a4a70>

x[[3]] <- 4
#> tracemem[0x55eec7b3af38 -> 0x55eec774cc18]:
```

We can avoid the copy by sub-assigning an integer instead of a double:

```
x <- c(1L, 2L, 3L)
tracemem(x)
#> <0x55eec6940ae0>

x[[3]] <- 4L
```

Please be aware that running this code in RStudio will result in additional copies because of the reference from the environment pane.

Q3: Sketch out the relationship between the following objects:

```
a <- 1:10
b <- list(a, a)
c <- list(b, a, 1:10)
```

A: `a` contains a reference to an address with the value `1:10`. `b` contains a list of two references to the same address as `a`. `c` contains a list of `b` (containing two references to `a`), `a` (containing the same reference again) and a reference pointing to a different address containing the same value (`1:10`).

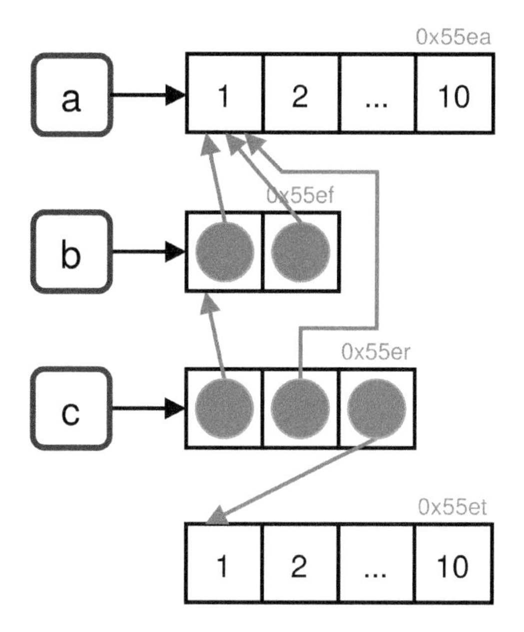

We can confirm these relationships by inspecting the reference tree in R.

```
ref(c)
#> █ [1:0x55erc93cbdd8] <list>    # c
#> ├─█ [2:0x55efcb8246e8] <list>  # - b
#> │ ├─[3:0x55eac7df4e98] <int>   # -- a
#> │ └─[3:0x55eac7df4e98]         # -- a
#> ├─[3:0x55eac7df4e98]           # - a
#> └─[4:0x55etc7aa6968] <int>     # - 1:10
```

Q4: What happens when you run this code:

```
x <- list(1:10)
x[[2]] <- x
```

Draw a picture.

A: The initial reference tree of `x` shows that the name `x` binds to a list object. This object contains a reference to the integer vector `1:10`.

```
x <- list(1:10)
```

```
ref(x)
#> █ [1:0x55853b74ff40] <list>
#> └─[2:0x534t3abffad8] <int>
```

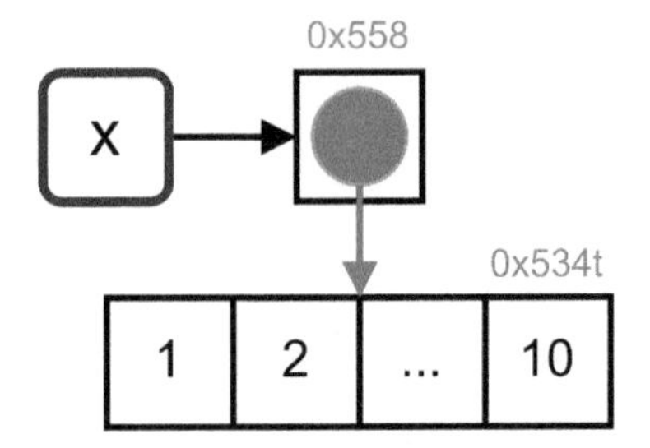

When x is assigned to an element of itself, copy-on-modify takes place and the list is copied to a new address in memory.

```
tracemem(x)
x[[2]] <- x
#> tracemem[0x55853b74ff40 -> 0x5d553bacdcd8]:
```

The list object previously bound to x is now referenced in the newly created list object. It is no longer bound to a name. The integer vector is referenced twice.

```
ref(x)
#> █ [1:0x5d553bacdcd8] <list>
#> ├─[2:0x534t3abffad8] <int>
#> └─█ [3:0x55853b74ff40] <list>
#>   └─[2:0x534t3abffad8]
```

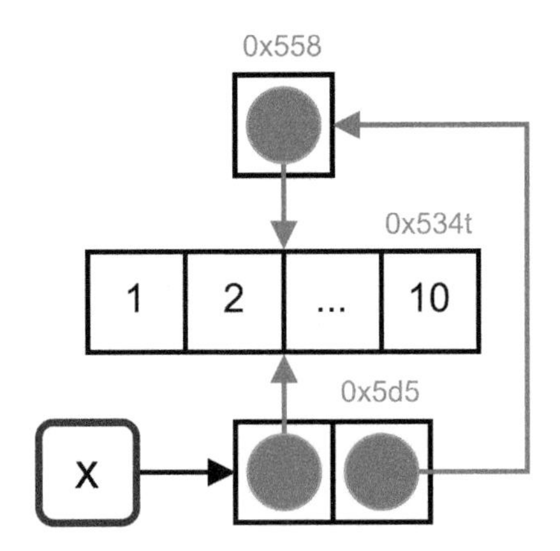

2.4 Object size

Q1: In the following example, why are `object.size(y)` and `obj_size(y)` so radically different? Consult the documentation of `object.size()`.

```
y <- rep(list(runif(1e4)), 100)

object.size(y)
#> 8005648 bytes
obj_size(y)
#> 80,896 B
```

A: `object.size()` doesn't account for shared elements within lists. Therefore, the results differ by a factor of ~ 100.

Q2: Take the following list. Why is its size somewhat misleading?

```
funs <- list(mean, sd, var)
obj_size(funs)
#> 17,608 B
```

A: All three functions are built-in to R as part of the `{base}` and `{stats}` packages and hence always available. So, what does it mean to measure the size of something that's already included in R?

(There's typically a more general question about what you want to know when you ask for the size of something — do you want to know how much data you'd need to send to communicate the object to someone else (e.g. serialise it), or do you want to know how much memory you'd free if you deleted it?)

Let us look for how many other objects this applies to as well.

The following packages are usually loaded by default.

```
base_pkgs <- c(
  "package:stats", "package:graphics", "package:grDevices",
  "package:utils", "package:datasets", "package:methods",
  "package:base"
)
```

To look up all functions from these packages we iterate over `base_pkgs` and apply `ls()` and `mget()` within each iteration.

```
base_objs <- base_pkgs %>%
  lapply(as.environment) %>%
  lapply(function(x) mget(ls(x, all.names = TRUE), x)) %>%
  setNames(base_pkgs)
```

This gives us more than 2700 objects which are usually available by default.

```
sum(lengths(base_objs))
#> [1] 2709

# We can also show the sizes in MB per package
vapply(base_objs, obj_size, double(1)) / 1024^2
#>     package:stats  package:graphics package:grDevices
#>             11.00              3.08              1.97
#>     package:utils  package:datasets   package:methods
#>              7.09              0.54             13.23
#>      package:base
#>             18.85

# Check if we've over-counted
as.numeric(obj_size(!!!base_objs)) / 1024^2
#> [1] 54
```

Q3: Predict the output of the following code:

```
a <- runif(1e6)
obj_size(a)

b <- list(a, a)
obj_size(b)
obj_size(a, b)

b[[1]][[1]] <- 10
obj_size(b)
obj_size(a, b)

b[[2]][[1]] <- 10
obj_size(b)
obj_size(a, b)
```

A: In R (on most platforms) a length-0 vector has 48 bytes of overhead.

```
obj_size(list())
#> 48 B
obj_size(double())
#> 48 B
obj_size(character())
#> 48 B
```

A single double takes up an additional 8 bytes of memory.

```
obj_size(double(1))
#> 56 B
obj_size(double(2))
#> 64 B
```

So, a 1 million double should take up 8,000,048 bytes.

```
a <- runif(1e6)
obj_size(a)
#> 8,000,048 B
```

(If you look carefully at the amount of memory occupied by short vectors, you will notice that the pattern is actually more complicated. This has to do with how R allocates memory and is not that important. If you want to know the full details, they're discussed in the 1st edition of *Advanced R*: `http://adv-r.had.co.nz/memory.html#object-size`).

For `b <- list(a, a)` both list elements contain references to the same memory address.

```
b <- list(a, a)
ref(a, b)
#> [1:0x5639afc25ed0] <dbl>
#>
#> █ [2:0x5639a8f3ef08] <list>
#> ├─[1:0x5639afc25ed0]
#> └─[1:0x5639afc25ed0]
```

Therefore, no additional memory is required for the second list element. The list itself requires 64 bytes, 48 bytes for an empty list and 8 bytes for each element (`obj_size(vector("list", 2))`). This lets us predict 8,000,048 B + 64 B = 8,000,112 B.

```
obj_size(b)
#> 8,000,112 B
```

When we modify the first element of `b[[1]]` copy-on-modify occurs. Both elements will still have the same size (8,000,040 B), but the first one gets a new address in memory. As `b`'s elements don't share references anymore, its object size adds up to the sum of the elements and the length-2 list: 8,000,048 B + 8,000,048 B + 64 B = 16,000,160 B (16 MB).

```
b[[1]][[1]] <- 10
obj_size(b)
#> 16,000,160 B
```

The second element of `b` still references the same address as `a`, so the combined size of `a` and `b` is the same as `b`.

```
obj_size(a, b)
#> 16,000,160 B
ref(a, b)
#> [1:0x5639afc25ed0] <dbl>
#>
#> █ [2:0x5639adf46fa8] <list>
#> ├─[3:0x5639b2c737c0] <dbl>
#> └─[1:0x5639afc25ed0]
```

When we modify the second element of `b`, this element will also point to a new memory address. This does not affect the size of the list.

```
b[[2]][[1]] <- 10
obj_size(b)
#> 16,000,160 B
```

However, as `b` doesn't share references with `a` anymore, the memory usage of the combined objects increases.

```
ref(a, b)
#> [1:0x5639afc25ed0] <dbl>
#>
#> █ [2:0x5639ad94c2c8] <list>
#> ├─[3:0x5639b2c737c0] <dbl>
#> └─[4:0x5639b0deb050] <dbl>
```

```
obj_size(a, b)
#> 24,000,208 B
```

2.5 Modify-in-place

Q1: Explain why the following code doesn't create a circular list.

```
x <- list()
x[[1]] <- x
```

A: In this situation copy-on-modify prevents the creation of a circular list. Let us step through the details:

```
x <- list()  # creates initial object
obj_addr(x)
#> [1] "0x55862f23ab80"

tracemem(x)
#> [1] "<0x55862f23ab80>"
x[[1]] <- x  # Copy-on-modify triggers new copy
#> tracemem[0x55862f23ab80 -> 0x55862e8ce028]:

obj_addr(x)       # copied object has new memory address
#> [1] "0x55862e8ce028"
obj_addr(x[[1]])  # list element contains old memory address
#> [1] "0x55862f23ab80"
```

Q2: Wrap the two methods for subtracting medians into two functions, then use the `{bench}` package to carefully compare their speeds. How does performance change as the number of columns increase?

A: First, we define a function to create some random data.

```
create_random_df <- function(nrow, ncol) {
  random_matrix <- matrix(runif(nrow * ncol), nrow = nrow)
  as.data.frame(random_matrix)
}
```

```
create_random_df(2, 2)
#>       V1     V2
#> 1 0.972 0.0116
#> 2 0.849 0.4339
```

Next, we wrap the two approaches to subtract numerical values (in our case medians) from each column of a data frame in their own function. We name these functions depending on whether the approach operates on a data frame or a list. For a fairer comparison, the second function also contains the overhead code to convert between data frame and list objects.

```
subtract_df <- function(x, medians) {
  for (i in seq_along(medians)) {
    x[[i]] <- x[[i]] - medians[[i]]
  }
  x
}

subtract_list <- function(x, medians) {
  x <- as.list(x)
  x <- subtract_df(x, medians)
  list2DF(x)
}
```

This lets us profile the performance, via benchmarks on data frames with differing numbers of columns. Therefore, we create a small helper that creates our random data frame and its medians before it benchmarks the two approaches by employing the {bench} package [Hester, 2020].

```
benchmark_medians <- function(ncol) {
  df <- create_random_df(nrow = 1e4, ncol = ncol)
  medians <- vapply(df, median, numeric(1), USE.NAMES = FALSE)

  bench::mark(
    "data frame" = subtract_df(df, medians),
    "list" = subtract_list(df, medians),
    time_unit = "ms"
  )
}

benchmark_medians(1)
#> # A tibble: 2 x 6
#>   expression    min median `itr/sec` mem_alloc `gc/sec`
```

```
#>    <bch:expr>  <dbl>  <dbl>     <dbl> <bch:byt>    <dbl>
#> 1 data frame 0.0240 0.0267    34009.     344KB     34.0
#> 2 list       0.0387 0.0443    21319.     156KB     43.3
```

The `bench::press()` function allows us to run our helper across a grid of parameters. We will use it to slowly increase the number of data frame columns in our benchmark.

```
results <- bench::press(
  ncol = c(1, 10, 50, 100, 250, 300, 400, 500, 750, 1000),
  benchmark_medians(ncol)
)
#> Running with:
#>     ncol
#>  1     1
#>  2    10
#>  3    50
#>  4   100
#>  5   250
#>  6   300
#>  7   400
#>  8   500
#>  9   750
#> 10  1000
```

Finally, we can plot and interpret our results.

```
library(ggplot2)

ggplot(
  results,
  aes(ncol, median, col = attr(expression, "description"))
) +
  geom_point(size = 2) +
  geom_smooth() +
  labs(
    x = "Number of Columns",
    y = "Execution Time (ms)",
    colour = "Data Structure"
  ) +
  theme(legend.position = "top")
#> `geom_smooth()` using method = 'loess' and formula 'y ~ x'
```

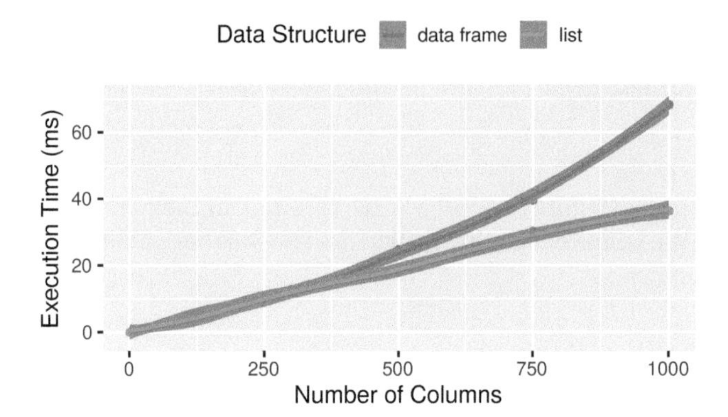

When working directly with the data frame, the execution time grows quadratically with the number of columns in the input data. This is because (e.g.) the first column must be copied n times, the second column n-1 times, and so on. When working with a list, the execution time increases only linearly.

Obviously in the long run, linear growth creates shorter run-times, but there is some cost to this strategy — we have to convert between data structures with `as.list()` and `list2DF()`. Even though this is fast and probably doesn't hurt much, the improved approach doesn't really pay off in this scenario until we get to a data frame that is about 300 columns wide (with the exact value depending on the characteristics of the system running the code).

Q3: What happens if you attempt to use `tracemem()` on an environment?

A: `tracemem()` cannot be used to mark and trace environments.

```
x <- new.env()
tracemem(x)
#> Error in tracemem(x): 'tracemem' is not useful for promise and
#> environment objects
```

The error occurs because "it is not useful to trace NULL, environments, promises, weak references, or external pointer objects, as these are not duplicated" (see `?tracemem`). Environments are always modified in place.

3

Vectors

3.2 Atomic vectors

Q1: How do you create raw and complex scalars? (See `?raw` and `?complex`.)

A: In R, scalars are represented as vectors of length one. However, there's no built-in syntax like there is for logicals, integers, doubles, and character vectors to create individual raw and complex values. Instead, you have to create them by calling a function.

For raw vectors you can use either `as.raw()` or `charToRaw()` to create them from numeric or character values.

```
as.raw(42)
#> [1] 2a
charToRaw("A")
#> [1] 41
```

In the case of complex numbers, real and imaginary parts may be provided directly to the `complex()` constructor.

```
complex(length.out = 1, real = 1, imaginary = 1)
#> [1] 1+1i
```

You can create purely imaginary numbers (e.g.) `1i`, but there is no way to create complex numbers without `+` (e.g. `1i + 1`).

Q2: Test your knowledge of vector coercion rules by predicting the output of the following uses of `c()`:

```
c(1, FALSE)      # will be coerced to double    -> 1 0
c("a", 1)        # will be coerced to character -> "a" "1"
c(TRUE, 1L)      # will be coerced to integer   -> 1 1
```

Q3: Why is `1 == "1"` true? Why is `-1 < FALSE` true? Why is `"one" < 2` false?

DOI: 10.1201/9781003175414-3

A: These comparisons are carried out by operator-functions (==, <), which coerce their arguments to a common type. In the examples above, these types will be character, double and character: `1` will be coerced to `"1"`, `FALSE` is represented as `0` and `2` turns into `"2"` (and numbers precede letters in lexicographic order (may depend on locale)).

Q4: Why is the default missing value, `NA`, a logical vector? What's special about logical vectors? (Hint: think about `c(FALSE, NA_character_)`.)

A: The presence of missing values shouldn't affect the type of an object. Recall that there is a type-hierarchy for coercion from character → double → integer → logical. When combining `NA`s with other atomic types, the `NA`s will be coerced to integer (`NA_integer_`), double (`NA_real_`) or character (`NA_character_`) and not the other way round. If `NA` were a character and added to a set of other values all of these would be coerced to character as well.

Q5: Precisely what do `is.atomic()`, `is.numeric()`, and `is.vector()` test for?

A: The documentation states that:

- `is.atomic()` tests if an object is an atomic vector (as defined in *Advanced R*) or is `NULL` (!).
- `is.numeric()` tests if an object has type integer or double and is not of class `factor`, `Date`, `POSIXt` or `difftime`.
- `is.vector()` tests if an object is a vector (as defined in *Advanced R*) or an expression and has no attributes, apart from names.

Atomic vectors are defined in *Advanced R* as objects of type logical, integer, double, complex, character or raw. Vectors are defined as atomic vectors or lists.

3.3 Attributes

Q1: How is `setNames()` implemented? How is `unname()` implemented? Read the source code.

A: `setNames()` is implemented as:

```
setNames <- function(object = nm, nm) {
  names(object) <- nm
  object
}
```

Because the data argument comes first, `setNames()` also works well with the magrittr-pipe operator. When no first argument is given, the result is a named vector (this is rather untypical as required arguments usually come first):

```
setNames( , c("a", "b", "c"))
#>   a   b   c
#> "a" "b" "c"
```

`unname()` is implemented in the following way:

```
unname <- function(obj, force = FALSE) {
  if (!is.null(names(obj)))
    names(obj) <- NULL
  if (!is.null(dimnames(obj)) && (force || !is.data.frame(obj)))
    dimnames(obj) <- NULL
  obj
}
```

`unname()` removes existing names (or dimnames) by setting them to `NULL`.

Q2: What does `dim()` return when applied to a 1-dimensional vector? When might you use `NROW()` or `NCOL()`?

A: From `?nrow`:

> `dim()` will return `NULL` when applied to a 1d vector.

One may want to use `NROW()` or `NCOL()` to handle atomic vectors, lists and NULL values in the same way as one column matrices or data frames. For these objects `nrow()` and `ncol()` return `NULL`:

```
x <- 1:10

# Return NULL
nrow(x)
#> NULL
ncol(x)
#> NULL

# Pretend it's a column vector
NROW(x)
#> [1] 10
NCOL(x)
#> [1] 1
```

Q3: How would you describe the following three objects? What makes them different to `1:5`?

```
x1 <- array(1:5, c(1, 1, 5))  # 1 row,  1 column,  5 in third dim.
x2 <- array(1:5, c(1, 5, 1))  # 1 row,  5 columns, 1 in third dim.
x3 <- array(1:5, c(5, 1, 1))  # 5 rows, 1 column,  1 in third dim.
```

A: These are all "one dimensional". If you imagine a 3d cube, `x1` is in the x-dimension, `x2` is in the y-dimension, and `x3` is in the z-dimension. In contrast to `1:5`, `x1`, `x2` and `x3` have a `dim` attribute.

Q4: An early draft used this code to illustrate `structure()`:

```
structure(1:5, comment = "my attribute")
#> [1] 1 2 3 4 5
```

But when you print that object you don't see the comment attribute. Why? Is the attribute missing, or is there something else special about it? (Hint: try using help.)

A: The documentation states (see `?comment`):

> Contrary to other attributes, the comment is not printed (by print or print.default).

Also, from `?attributes`:

> Note that some attributes (namely class, comment, dim, dimnames, names, row.names and tsp) are treated specially and have restrictions on the values which can be set.

We can retrieve comment attributes by calling them explicitly:

```
foo <- structure(1:5, comment = "my attribute")

attributes(foo)
#> $comment
#> [1] "my attribute"
attr(foo, which = "comment")
#> [1] "my attribute"
```

3.4 S3 atomic vectors

Q1: What sort of object does `table()` return? What is its type? What attributes does it have? How does the dimensionality change as you tabulate more variables?

A: `table()` returns a contingency table of its input variables. It is implemented as an integer vector with class `table` and dimensions (which makes it act like an array). Its attributes are `dim` (dimensions) and `dimnames` (one name for each input column). The dimensions correspond to the number of unique values (factor levels) in each input variable.

```
x <- table(mtcars[c("vs", "cyl", "am")])

typeof(x)
#> [1] "integer"
attributes(x)
#> $dim
#> [1] 2 3 2
#>
#> $dimnames
#> $dimnames$vs
#> [1] "0" "1"
#>
#> $dimnames$cyl
#> [1] "4" "6" "8"
#>
#> $dimnames$am
#> [1] "0" "1"
#>
#>
#> $class
#> [1] "table"

# Subset x like it's an array
x[ , , 1]
#>    cyl
#> vs   4  6  8
#>   0  0  0 12
#>   1  3  4  0
x[ , , 2]
#>    cyl
```

```
#> vs  4 6 8
#>   0 1 3 2
#>   1 7 0 0
```

Q2: What happens to a factor when you modify its levels?

```
f1 <- factor(letters)
levels(f1) <- rev(levels(f1))
```

A: The underlying integer values stay the same, but the levels are changed, making it look like the data has changed.

```
f1 <- factor(letters)
f1
#>  [1] a b c d e f g h i j k l m n o p q r s t u v w x y z
#> Levels: a b c d e f g h i j k l m n o p q r s t u v w x y z
as.integer(f1)
#>  [1]  1  2  3  4  5  6  7  8  9 10 11 12 13 14 15 16 17 18 19 20 21
#> [22] 22 23 24 25 26

levels(f1) <- rev(levels(f1))
f1
#>  [1] z y x w v u t s r q p o n m l k j i h g f e d c b a
#> Levels: z y x w v u t s r q p o n m l k j i h g f e d c b a
as.integer(f1)
#>  [1]  1  2  3  4  5  6  7  8  9 10 11 12 13 14 15 16 17 18 19 20 21
#> [22] 22 23 24 25 26
```

Q3: What does this code do? How do `f2` and `f3` differ from `f1`?

```
f2 <- rev(factor(letters))

f3 <- factor(letters, levels = rev(letters))
```

A: For `f2` and `f3` either the order of the factor elements *or* its levels are being reversed. For `f1` both transformations are occurring.

```
# Reverse element order
(f2 <- rev(factor(letters)))
#>  [1] z y x w v u t s r q p o n m l k j i h g f e d c b a
#> Levels: a b c d e f g h i j k l m n o p q r s t u v w x y z
as.integer(f2)
```

```
#>  [1] 26 25 24 23 22 21 20 19 18 17 16 15 14 13 12 11 10  9  8  7  6
#> [22]  5  4  3  2  1

# Reverse factor levels (when creating factor)
(f3 <- factor(letters, levels = rev(letters)))
#>  [1] a b c d e f g h i j k l m n o p q r s t u v w x y z
#> Levels: z y x w v u t s r q p o n m l k j i h g f e d c b a
as.integer(f3)
#>  [1] 26 25 24 23 22 21 20 19 18 17 16 15 14 13 12 11 10  9  8  7  6
#> [22]  5  4  3  2  1
```

3.5 Lists

Q1: List all the ways that a list differs from an atomic vector.

A: To summarise:

- Atomic vectors are always homogeneous (all elements must be of the same type). Lists may be heterogeneous (the elements can be of different types) as described in the introduction of the vectors chapter (`https://adv-r.hadley.nz/vectors-chap.html#introduction`).
- Atomic vectors point to one address in memory, while lists contain a separate reference for each element. (This was described in the list sections of the vectors (`https://adv-r.hadley.nz/vectors-chap.html#lists`) and the names and values (`https://adv-r.hadley.nz/names-values.html#list-references`) chapters.)

```
lobstr::ref(1:2)
#> [1:0x55ff5ba8d310] <int>
lobstr::ref(list(1:2, 2))
#> █ [1:0x55ff5dd10ff8] <list>
#> ├─[2:0x55ff5c3cdcc8] <int>
#> └─[3:0x55ff5ec90ea0] <dbl>
```

- Subsetting with out-of-bounds and `NA` values leads to different output. For example, `[` returns `NA` for atomics and `NULL` for lists. (This is described in more detail within the subsetting chapter (`https://adv-r.hadley.nz/subsetting.html`).)

```
# Subsetting atomic vectors
(1:2)[3]
#> [1] NA
(1:2)[NA]
#> [1] NA NA

# Subsetting lists
as.list(1:2)[3]
#> [[1]]
#> NULL
as.list(1:2)[NA]
#> [[1]]
#> NULL
#>
#> [[2]]
#> NULL
```

Q2: Why do you need to use `unlist()` to convert a list to an atomic vector? Why doesn't `as.vector()` work?

A: A list is already a vector, though not an atomic one!

Note that `as.vector()` and `is.vector()` use different definitions of "vector"!

```
is.vector(as.vector(mtcars))
#> [1] FALSE
```

Q3: Compare and contrast `c()` and `unlist()` when combining a date and date-time into a single vector.

A: Date and date-time objects are both built upon doubles. While dates store the number of days since the reference date 1970-01-01 (also known as "the Epoch") in days, date-time-objects (POSIXct) store the time difference to this date in seconds.

```
date    <- as.Date("1970-01-02")
dttm_ct <- as.POSIXct("1970-01-01 01:00", tz = "UTC")

# Internal representations
unclass(date)
#> [1] 1
unclass(dttm_ct)
#> [1] 3600
#> attr(,"tzone")
#> [1] "UTC"
```

As the `c()` generic only dispatches on its first argument, combining date and date-time objects via `c()` could lead to surprising results in older R versions (pre R 4.0.0):

```
# Output in R version 3.6.2
c(date, dttm_ct) # equal to c.Date(date, dttm_ct)
#> [1] "1970-01-02" "1979-11-10"
c(dttm_ct, date) # equal to c.POSIXct(date, dttm_ct)
#> [1] "1970-01-01 02:00:00 CET" "1970-01-01 01:00:01 CET"
```

In the first statement above `c.Date()` is executed, which incorrectly treats the underlying double of `dttm_ct` (3600) as days instead of seconds. Conversely, when `c.POSIXct()` is called on a date, one day is counted as one second only.

We can highlight these mechanics by the following code:

```
# Output in R version 3.6.2
unclass(c(date, dttm_ct)) # internal representation
#> [1] 1 3600
date + 3599
#> "1979-11-10"
```

As of R 4.0.0 these issues have been resolved and both methods now convert their input first into `POSIXct` and `Date`, respectively.

```
c(dttm_ct, date)
#> [1] "1970-01-01 02:00:00 CET" "1970-01-02 01:00:00 CET"
unclass(c(dttm_ct, date))
#> [1]  3600 86400

c(date, dttm_ct)
#> [1] "1970-01-02" "1970-01-01"
unclass(c(date, dttm_ct))
#> [1] 1 0
```

However, as `c()` strips the time zone (and other attributes) of `POSIXct` objects, some caution is still recommended.

```
(dttm_ct <- as.POSIXct("1970-01-01 01:00", tz = "HST"))
#> [1] "1970-01-01 01:00:00 HST"
attributes(c(dttm_ct))
```

```
#> $class
#> [1] "POSIXct" "POSIXt"
```

A package that deals with these kinds of problems in more depth and provides a structural solution for them is the `{vctrs}` package (`https://github.com/r-lib/vctrs`) [Wickham et al., 2020b] which is also used throughout the tidyverse [Wickham et al., 2019].

Let's look at `unlist()`, which operates on list input.

```
# Attributes are stripped
unlist(list(date, dttm_ct))
#> [1]     1 39600
```

We see again that dates and date-times are internally stored as doubles. Unfortunately, this is all we are left with, when unlist strips the attributes of the list.

To summarise: `c()` coerces types and strips time zones. Errors may have occurred in older R versions because of inappropriate method dispatch/immature methods. `unlist()` strips attributes.

3.6 Data frames and tibbles

Q1: Can you have a data frame with zero rows? What about zero columns?

A: Yes, you can create these data frames easily; either during creation or via subsetting. Even both dimensions can be zero.

Create a 0-row, 0-column, or an empty data frame directly:

```
data.frame(a = integer(), b = logical())
#> [1] a b
#> <0 rows> (or 0-length row.names)

data.frame(row.names = 1:3)  # or data.frame()[1:3, ]
#> data frame with 0 columns and 3 rows

data.frame()
#> data frame with 0 columns and 0 rows
```

Create similar data frames via subsetting the respective dimension with either `0`, `NULL`, `FALSE` or a valid 0-length atomic (`logical(0)`, `character(0)`, `integer(0)`, `double(0)`). Negative integer sequences would also work. The following example uses a zero:

```
mtcars[0, ]
#>  [1] mpg  cyl  disp hp   drat wt   qsec vs   am   gear carb
#> <0 rows> (or 0-length row.names)

mtcars[ , 0]  # or mtcars[0]
#> data frame with 0 columns and 32 rows

mtcars[0, 0]
#> data frame with 0 columns and 0 rows
```

Q2: What happens if you attempt to set rownames that are not unique?

A: Matrices can have duplicated row names, so this does not cause problems.

Data frames, however, require unique rownames and you get different results depending on how you attempt to set them. If you set them directly or via `row.names()`, you get an error:

```
data.frame(row.names = c("x", "y", "y"))
#> Error in data.frame(row.names = c("x", "y", "y")): duplicate
#> row.names: y

df <- data.frame(x = 1:3)
row.names(df) <- c("x", "y", "y")
#> Warning: non-unique value when setting 'row.names': 'y'
#> Error in `.rowNamesDF<-`(x, value = value): duplicate 'row.names'
#> are not allowed
```

If you use subsetting, `[` automatically deduplicates:

```
row.names(df) <- c("x", "y", "z")
df[c(1, 1, 1), , drop = FALSE]
#>     x
#> x   1
#> x.1 1
#> x.2 1
```

Q3: If `df` is a data frame, what can you say about `t(df)`, and `t(t(df))`? Perform some experiments, making sure to try different column types.

A: Both of `t(df)` and `t(t(df))` will return matrices:

```
df <- data.frame(x = 1:3, y = letters[1:3])
is.matrix(df)
#> [1] FALSE
is.matrix(t(df))
#> [1] TRUE
is.matrix(t(t(df)))
#> [1] TRUE
```

The dimensions will respect the typical transposition rules:

```
dim(df)
#> [1] 3 2
dim(t(df))
#> [1] 2 3
dim(t(t(df)))
#> [1] 3 2
```

Because the output is a matrix, every column is coerced to the same type. (It is implemented within `t.data.frame()` via `as.matrix()` which is described below).

```
df
#>   x y
#> 1 1 a
#> 2 2 b
#> 3 3 c
t(df)
#>   [,1] [,2] [,3]
#> x "1"  "2"  "3"
#> y "a"  "b"  "c"
```

Q4: What does `as.matrix()` do when applied to a data frame with columns of different types? How does it differ from `data.matrix()`?

A: The type of the result of `as.matrix` depends on the types of the input columns (see `?as.matrix`):

> The method for data frames will return a character matrix if there is only atomic columns and any non-(numeric/logical/complex) column, applying as.vector to factors and format to other non-character columns. Otherwise the usual coercion hierarchy (logical < integer < double < complex) will be used, e.g. all-logical data frames will be coerced to a logical matrix, mixed logical-integer will give an integer matrix, etc.

On the other hand, `data.matrix` will always return a numeric matrix (see `?data.matrix()`).

> Return the matrix obtained by converting all the variables in a data frame to numeric mode and then binding them together as the columns of a matrix. Factors and ordered factors are replaced by their internal codes. [...] Character columns are first converted to factors and then to integers.

We can illustrate and compare the mechanics of these functions using a concrete example. `as.matrix()` makes it possible to retrieve most of the original information from the data frame but leaves us with characters. To retrieve all information from `data.matrix()`'s output, we would need a lookup table for each column.

```
df_coltypes <- data.frame(
  a = c("a", "b"),
  b = c(TRUE, FALSE),
  c = c(1L, 0L),
  d = c(1.5, 2),
  e = factor(c("f1", "f2"))
)

as.matrix(df_coltypes)
#>      a   b       c   d     e
#> [1,] "a" "TRUE"  "1" "1.5" "f1"
#> [2,] "b" "FALSE" "0" "2.0" "f2"
data.matrix(df_coltypes)
#>      a b c   d e
#> [1,] 1 1 1 1.5 1
#> [2,] 2 0 0 2.0 2
```

4

Subsetting

4.2 Selecting multiple elements

Q1: Fix each of the following common data frame subsetting errors:

```
mtcars[mtcars$cyl = 4, ]
# use `==`              (instead of `=`)

mtcars[-1:4, ]
# use `-(1:4)`          (instead of `-1:4`)

mtcars[mtcars$cyl <= 5]
# `,` is missing

mtcars[mtcars$cyl == 4 | 6, ]
# use `mtcars$cyl == 6` (instead of `6`)
#  or `%in% c(4, 6)`    (instead of `== 4 | 6`)
```

Q2: Why does the following code yield five missing values? (Hint: why is it different from `x[NA_real_]`?)

```
x <- 1:5
x[NA]
#> [1] NA NA NA NA NA
```

A: In contrast to `NA_real`, `NA` has logical type and logical vectors are recycled to the same length as the vector being subset, i.e. `x[NA]` is recycled to `x[NA, NA, NA, NA, NA]`.

Q3: What does `upper.tri()` return? How does subsetting a matrix with it work? Do we need any additional subsetting rules to describe its behaviour?

```
x <- outer(1:5, 1:5, FUN = "*")
x[upper.tri(x)]
```

DOI: 10.1201/9781003175414-4

A: `upper.tri(x)` returns a logical matrix, which contains `TRUE` values above the diagonal and `FALSE` values everywhere else. In `upper.tri()` the positions for `TRUE` and `FALSE` values are determined by comparing x's row and column indices via `.row(dim(x)) < .col(dim(x))`.

```
x
#>      [,1] [,2] [,3] [,4] [,5]
#> [1,]    1    2    3    4    5
#> [2,]    2    4    6    8   10
#> [3,]    3    6    9   12   15
#> [4,]    4    8   12   16   20
#> [5,]    5   10   15   20   25
upper.tri(x)
#>       [,1]  [,2]  [,3]  [,4]  [,5]
#> [1,] FALSE  TRUE  TRUE  TRUE  TRUE
#> [2,] FALSE FALSE  TRUE  TRUE  TRUE
#> [3,] FALSE FALSE FALSE  TRUE  TRUE
#> [4,] FALSE FALSE FALSE FALSE  TRUE
#> [5,] FALSE FALSE FALSE FALSE FALSE
```

When subsetting with logical matrices, all elements that correspond to `TRUE` will be selected. Matrices extend vectors with a dimension attribute, so the vector forms of subsetting can be used (including logical subsetting). We should take care, that the dimensions of the subsetting matrix match the object of interest — otherwise unintended selections due to vector recycling may occur. Please also note, that this form of subsetting returns a vector instead of a matrix, as the subsetting alters the dimensions of the object.

```
x[upper.tri(x)]
#>  [1]  2  3  6  4  8 12  5 10 15 20
```

Q4: Why does `mtcars[1:20]` return an error? How does it differ from the similar `mtcars[1:20, ]`?

A: When subsetting a data frame with a single vector, it behaves the same way as subsetting a list of columns. So, `mtcars[1:20]` would return a data frame containing the first 20 columns of the dataset. However, as `mtcars` has only 11 columns, the index will be out of bounds and an error is thrown. `mtcars[1:20, ]` is subsetted with two vectors, so 2d subsetting kicks in, and the first index refers to rows.

Q5: Implement your own function that extracts the diagonal entries from a matrix (it should behave like `diag(x)` where `x` is a matrix).

A: The elements in the diagonal of a matrix have the same row- and column indices. This characteristic can be used to create a suitable numeric matrix used for subsetting.

```
diag2 <- function(x) {
  n <- min(nrow(x), ncol(x))
  idx <- cbind(seq_len(n), seq_len(n))

  x[idx]
}

# Let's check if it works
(x <- matrix(1:30, 5))
#>      [,1] [,2] [,3] [,4] [,5] [,6]
#> [1,]    1    6   11   16   21   26
#> [2,]    2    7   12   17   22   27
#> [3,]    3    8   13   18   23   28
#> [4,]    4    9   14   19   24   29
#> [5,]    5   10   15   20   25   30

diag(x)
#> [1]  1  7 13 19 25
diag2(x)
#> [1]  1  7 13 19 25
```

Q6: What does `df[is.na(df)] <- 0` do? How does it work?

A: This expression replaces the `NA`s in `df` with `0`. Here `is.na(df)` returns a logical matrix that encodes the position of the missing values in `df`. Subsetting and assignment are then combined to replace only the missing values.

4.3 Selecting a single element

Q1: Brainstorm as many ways as possible to extract the third value from the `cyl` variable in the `mtcars` dataset.

A: Base R already provides an abundance of possibilities:

```
# Select column first
mtcars$cyl[[3]]
#> [1] 4
```

```
mtcars[ , "cyl"][[3]]
#> [1] 4
mtcars[["cyl"]][[3]]
#> [1] 4
with(mtcars, cyl[[3]])
#> [1] 4

# Select row first
mtcars[3, ]$cyl
#> [1] 4
mtcars[3, "cyl"]
#> [1] 4
mtcars[3, ][ , "cyl"]
#> [1] 4
mtcars[3, ][["cyl"]]
#> [1] 4

# Select simultaneously
mtcars[3, 2]
#> [1] 4
mtcars[[c(2, 3)]]
#> [1] 4
```

Q2: Given a linear model, e.g. `mod <- lm(mpg ~ wt, data = mtcars)`, extract the residual degrees of freedom. Extract the R squared from the model summary (`summary(mod)`).

A: `mod` is of type list, which opens up several possibilities. We use `$` or `[[` to extract a single element:

```
mod <- lm(mpg ~ wt, data = mtcars)

mod$df.residual
#> [1] 30
mod[["df.residual"]]
#> [1] 30
```

The same also applies to `summary(mod)`, so we could use, e.g.:

```
summary(mod)$r.squared
#> [1] 0.753
```

(Tip: The {broom} package (https://github.com/tidymodels/broom) [Robinson et al., 2020] provides a very useful approach to work with models in a tidy way.)

4.5 Applications

Q1: How would you randomly permute the columns of a data frame? (This is an important technique in random forests.) Can you simultaneously permute the rows and columns in one step?

A: This can be achieved by combining `[` and `sample()`:

```
# Permute columns
mtcars[sample(ncol(mtcars))]

# Permute columns and rows in one step
mtcars[sample(nrow(mtcars)), sample(ncol(mtcars))]
```

Q2: How would you select a random sample of `m` rows from a data frame? What if the sample had to be contiguous (i.e. with an initial row, a final row, and every row in between)?

A: Selecting `m` random rows from a data frame can be achieved through subsetting.

```
m <- 10
mtcars[sample(nrow(mtcars), m), ]
```

Holding successive lines together as a blocked sample requires only a certain amount of caution in order to obtain the correct start and end index.

```
start <- sample(nrow(mtcars) - m + 1, 1)
end <- start + m - 1
mtcars[start:end, , drop = FALSE]
```

Q3: How could you put the columns in a data frame in alphabetical order?

A: We combine `[` with `order()` or `sort()`:

```
mtcars[order(names(mtcars))]
mtcars[sort(names(mtcars))]
```

5

Control flow

5.2 Choices

Q1: What type of vector does each of the following calls to `ifelse()` return?

```
ifelse(TRUE, 1, "no")
ifelse(FALSE, 1, "no")
ifelse(NA, 1, "no")
```

Read the documentation and write down the rules in your own words.

A: The arguments of `ifelse()` are named `test`, `yes` and `no`. In general, `ifelse()` returns the entry for `yes` when `test` is `TRUE`, the entry for `no` when `test` is `FALSE` and `NA` when `test` is `NA`. Therefore, the expressions above return vectors of type `double` (`1`), `character` (`"no"`) and `logical` (`NA`).

To be a little more precise, we will cite the part of the documentation on the return value of `ifelse()`:

> A vector of the same length and attributes (including dimensions and "class") as test and data values from the values of yes or no. The mode of the answer will be coerced from logical to accommodate first any values taken from yes and then any values taken from no.

This is surprising because it uses the type of `test`. In practice this means, that `test` is first converted to logical and if the result is neither `TRUE` nor `FALSE`, simply `as.logical(test)` is returned.

```
ifelse(logical(), 1, "no")
#> logical(0)
ifelse(NaN, 1, "no")
#> [1] NA
ifelse(NA_character_, 1, "no")
#> [1] NA
ifelse("a", 1, "no")
```

DOI: 10.1201/9781003175414-5

```
#> [1] NA
ifelse("true", 1, "no")
#> [1] 1
```

Q2: Why does the following code work?

```
x <- 1:10
if (length(x)) "not empty" else "empty"
#> [1] "not empty"

x <- numeric()
if (length(x)) "not empty" else "empty"
#> [1] "empty"
```

A: `if()` expects a logical condition, but also accepts a numeric vector where `0` is treated as `FALSE` and all other numbers are treated as `TRUE`. Numerical missing values (including `NaN`) lead to an error in the same way that a logical missing, `NA`, does.

5.3 Loops

Q1: Why does this code succeed without errors or warnings?

```
x <- numeric()
out <- vector("list", length(x))
for (i in 1:length(x)) {
  out[i] <- x[i] ^ 2
}
out
```

A: This loop is a delicate issue, and we have to consider a few points to explain why it is evaluated without raising any errors or warnings.

The beginning of this code smell is the statement `1:length(x)` which creates the index of the for loop. As `x` has length 0 `1:length(x)` counts down from 1 to 0. This issue is typically avoided via usage of `seq_along(x)` or similar helpers which would just generate `integer(0)` in this case.

As we use `[<-` and `[` for indexing 0-length vectors at their first and zeroth position, we need to be aware of their subsetting behaviour for out-of-bounds and zero indices.

During the first iteration `x[1]` will generate an `NA` (out-of-bounds indexing for atomics). The resulting `NA` (from squaring) will be assigned to the empty length-1 list `out[1]` (out-of-bounds indexing for lists).

In the next iteration, `x[0]` will return `numeric(0)` (zero indexing for atomics). Again, squaring doesn't change the value and `numeric(0)` is assigned to `out[0]` (zero indexing for lists). Assigning a 0-length vector to a 0-length subset works but doesn't change the object.

Overall, the code works, because each step includes valid R operations (even though the result may not be what the user intended).

Q2: When the following code is evaluated, what can you say about the vector being iterated?

```
xs <- c(1, 2, 3)
for (x in xs) {
  xs <- c(xs, x * 2)
}
xs
#> [1] 1 2 3 2 4 6
```

A: In this loop `x` takes on the values of the initial `xs` (`1`, `2` and `3`), indicating that it is evaluated just once in the beginning of the loop, not after each iteration. (Otherwise, we would run into an infinite loop.)

Q3: What does the following code tell you about when the index is updated?

```
for (i in 1:3) {
  i <- i * 2
  print(i)
}
#> [1] 2
#> [1] 4
#> [1] 6
```

A: In a for loop the index is updated in the beginning of each iteration. Therefore, reassigning the index symbol during one iteration doesn't affect the following iterations. (Again, we would otherwise run into an infinite loop.)

6

Functions

6.2 Function fundamentals

Q1: Given a name, like "mean", `match.fun()` lets you find a function. Given a function, can you find its name? Why doesn't that make sense in R?

A: In R there is no one-to-one mapping between functions and names. A name always points to a single object, but an object may have zero, one or many names.

Let's look at an example:

```
function(x) sd(x) / mean(x)
#> function(x) sd(x) / mean(x)

f1 <- function(x) (x - min(x)) / (max(x) - min(x))
f2 <- f1
f3 <- f1
```

While the function in the first line is not bound to a name multiple names (`f1`, `f2` and `f3`) point to the second function. So, the main point is that the relation between name and object is only clearly defined in one direction.

Besides that, there are obviously ways to search for function names. However, to be sure to find the right one(s), you should not only compare the code (body) but also the arguments (formals) and the creation environment. As `formals()`, `body()` and `environment()` all return `NULL` for primitive functions, the easiest way to check if two functions are exactly equal is just to use `identical()`.

Q2: It's possible (although typically not useful) to call an anonymous function. Which of the two approaches below is correct? Why?

```
function(x) 3()
#> function(x) 3()
```

DOI: 10.1201/9781003175414-6

```
(function(x) 3)()
#> [1] 3
```

A: The second approach is correct.

The anonymous function `function(x) 3` is surrounded by a pair of parentheses before it is called by `()`. These extra parentheses separate the function call from the anonymous function's body. Without them a function with the invalid body `3()` is returned, which throws an error when we call it. This is easier to see if we name the function:

```
f <- function(x) 3()
f
#> function(x) 3()
f()
#> Error in f(): attempt to apply non-function
```

Q3: A good rule of thumb is that an anonymous function should fit on one line and shouldn't need to use `{}`. Review your code. Where could you have used an anonymous function instead of a named function? Where should you have used a named function instead of an anonymous function?

A: The use of anonymous functions allows concise and elegant code in certain situations. However, they miss a descriptive name and when re-reading the code, it can take a while to figure out what they do. That's why it's helpful to give long and complex functions a descriptive name. It may be worthwhile to take a look at your own projects or other people's code to reflect on this part of your coding style.

Q4: What function allows you to tell if an object is a function? What function allows you to tell if a function is a primitive function?

A: Use `is.function()` to test if an object is a function. Consider using `is.primitive()` to test specifically for primitive functions.

Q5: This code makes a list of all functions in the `{base}` package.

```
objs <- mget(ls("package:base", all = TRUE), inherits = TRUE)
funs <- Filter(is.function, objs)
```

Use it to answer the following questions:

a. Which base function has the most arguments?
b. How many base functions have no arguments? What's special about those functions?

c. How could you adapt the code to find all primitive functions?

A: Let's look at each sub-question separately:

a. To find the function with the most arguments, we first compute the length of `formals()`.

```
library(purrr)

n_args <- funs %>%
  map(formals) %>%
  map_int(length)
```

Then we sort `n_args` in decreasing order and look at its first entries.

```
n_args %>%
  sort(decreasing = TRUE) %>%
  head()
#> scan format.default source
#> 22 16 16
#> formatC library merge.data.frame
#> 15 13 13
```

b. We can further use `n_args` to find the number of functions with no arguments:

```
sum(n_args == 0)
#> [1] 248
```

However, this over counts because `formals()` returns `NULL` for primitive functions, and `length(NULL)` is 0. To fix this, we can first remove the primitive functions:

```
n_args2 <- funs %>%
  discard(is.primitive) %>%
  map(formals) %>%
  map_int(length)

sum(n_args2 == 0)
#> [1] 47
```

Indeed, most of the functions with no arguments are actually primitive functions.

c. To find all primitive functions, we can change the predicate in `Filter()` from `is.function()` to `is.primitive()`:

```
funs <- Filter(is.primitive, objs)
length(funs)
#> [1] 201
```

Q6: What are the three important components of a function?

A: These components are the function's `body()`, `formals()` and `environment()`. However, as mentioned in *Advanced R*:

> There is one exception to the rule that functions have three components. Primitive functions, like `sum()`, call C code directly with `.Primitive()` and contain no R code. Therefore, their `formals()`, `body()`, and `environment()` are all `NULL`.

Q7: When does printing a function not show what environment it was created in?

A: Primitive functions and functions created in the global environment do not print their environment.

6.4 Lexical scoping

Q1: What does the following code return? Why? Describe how each of the three `c`'s is interpreted.

```
c <- 10
c(c = c)
```

A: This code returns a named numeric vector of length one — with one element of the value `10` and the name `"c"`. The first `c` represents the `c()` function, the second `c` is interpreted as a (quoted) name and the third `c` as a value.

Q2: What are the four principles that govern how R looks for values?

A: R's lexical scoping (`https://adv-r.hadley.nz/functions.html#lexical-scoping`) rules are based on these four principles:

- Name masking (`https://adv-r.hadley.nz/functions.html#name-masking`)
- Functions vs. variables (`https://adv-r.hadley.nz/functions.html#functions-versus-variables`)
- A fresh start (`https://adv-r.hadley.nz/functions.html#fresh-start`)
- Dynamic lookup (`https://adv-r.hadley.nz/functions.html#dynamic-lookup`)

Q3: What does the following function return? Make a prediction before running the code yourself.

```
f <- function(x) {
  f <- function(x) {
    f <- function() {
      x ^ 2
    }
    f() + 1
  }
  f(x) * 2
}
f(10)
```

A: Within this nested function two more functions also named `f` are defined and called. Because the functions are each executed in their own environment R will look up and use the functions defined last in these environments. The innermost `f()` is called last, though it is the first function to return a value. Therefore, the order of the calculation passes "from the inside to the outside" and the function returns `((10 ^ 2) + 1) * 2`, i.e. 202.

6.5 Lazy evaluation

Q1: What important property of `&&` makes `x_ok()` work?

```
x_ok <- function(x) {
  !is.null(x) && length(x) == 1 && x > 0
}

x_ok(NULL)
#> [1] FALSE
x_ok(1)
#> [1] TRUE
```

```
x_ok(1:3)
#> [1] FALSE
```

What is different with this code? Why is this behaviour undesirable here?

```
x_ok <- function(x) {
  !is.null(x) & length(x) == 1 & x > 0
}

x_ok(NULL)
#> logical(0)
x_ok(1)
#> [1] TRUE
x_ok(1:3)
#> [1] FALSE FALSE FALSE
```

A: In summary: `&&` short-circuits which means that if the left-hand side is `FALSE` it doesn't evaluate the right-hand side (because it doesn't matter). Similarly, if the left-hand side of `||` is `TRUE` it doesn't evaluate the right-hand side.

We expect `x_ok()` to validate its input via certain criteria: it must not be `NULL`, have length `1` and be greater than `0`. Meaningful outcomes for this assertion will be `TRUE`, `FALSE` or `NA`. The desired behaviour is reached by combining the assertions through `&&` instead of `&`.

`&&` does not perform elementwise comparisons; instead it uses the first element of each value only. It also uses lazy evaluation, in the sense that evaluation "proceeds only until the result is determined" (from `?Logic`). This means that the RHS of `&&` won't be evaluated if the LHS already determines the outcome of the comparison (e.g. evaluate to `FALSE`). This behaviour is also known as "short-circuiting". For some situations (`x = 1`) both operators will lead to the same result. But this is not always the case. For `x = NULL`, the `&&`-operator will stop after the `!is.null` statement and return the result. The following conditions won't even be evaluated! (If the other conditions are also evaluated (by the use of `&`), the outcome would change. `NULL > 0` returns `logical(0)`, which is not helpful in this case.)

We can also see the difference in behaviour, when we set `x = 1:3`. The `&&`-operator returns the result from `length(x) == 1`, which is `FALSE`. Using `&` as the logical operator leads to the (vectorised) `x > 0` condition being evaluated and also returned.

Q2: What does this function return? Why? Which principle does it illustrate?

```
f2 <- function(x = z) {
  z <- 100
  x
}
f2()
```

A: The function returns 100. The default argument (`x = z`) gets lazily evaluated within the function environment when `x` gets accessed. At this time `z` has already been bound to the value `100`. The illustrated principle here is *lazy evaluation*.

Q3: What does this function return? Why? Which principle does it illustrate?

```
y <- 10
f1 <- function(x = {y <- 1; 2}, y = 0) {
  c(x, y)
}
f1()
y
```

A: The function returns `c(2, 1)` which is due to *name masking*. When `x` is accessed within `c()`, the promise `x = {y <- 1; 2}` is evaluated inside `f1()`'s environment. `y` gets bound to the value `1` and the return value of `{()` (`2`) gets assigned to `x`. When `y` gets accessed next within `c()`, it has already the value `1` and R doesn't need to look it up any further. Therefore, the promise `y = 0` won't be evaluated. Also, as `y` is assigned within `f1()`'s environment, the value of the global variable `y` is left untouched.

Q4: In `hist()`, the default value of `xlim` is `range(breaks)`, the default value for `breaks` is `"Sturges"`, and

```
range("Sturges")
#> [1] "Sturges" "Sturges"
```

Explain how `hist()` works to get a correct `xlim` value.

A: The `xlim` argument of `hist()` defines the range of the histogram's x-axis. In order to provide a valid axis `xlim` must contain a numeric vector of exactly two unique values. Consequently, for the default `xlim = range(breaks)`), `breaks` must evaluate to a vector with at least two unique values.

During execution `hist()` overwrites the `breaks` argument. The `breaks` argument is quite flexible and allows the users to provide the breakpoints directly or compute them in several ways. Therefore, the specific behaviour depends highly on the input. But `hist` ensures that `breaks` evaluates to a numeric vector containing at least two unique elements before `xlim` is computed.

Q5: Explain why this function works. Why is it confusing?

```
show_time <- function(x = stop("Error!")) {
  stop <- function(...) Sys.time()
  print(x)
}
show_time()
#> [1] "2020-12-27 23:30:10 CET"
```

A: Before `show_time()` accesses `x` (default `stop("Error")`), the `stop()` function is masked by `function(...) Sys.time()`. As default arguments are evaluated in the function environment, `print(x)` will be evaluated as `print(Sys.time())`.

This function is confusing because its behaviour changes when `x`'s value is supplied directly. Now the value from the calling environment will be used and the overwriting of `stop()` won't affect `x` anymore.

```
show_time(x = stop("Error!"))
#> Error in print(x): Error!
```

Q6: How many arguments are required when calling `library()`?

A: `library()` doesn't require any arguments. When called without arguments `library()` invisibly returns a list of class `libraryIQR`, which contains a results matrix with one row and three columns per installed package. These columns contain entries for the name of the package ("Package"), the path to the package ("LibPath") and the title of the package ("Title"). `library()` also has its own print method (`print.libraryIQR()`), which displays this information conveniently in its own window.

This behaviour is also documented under the details section of `library()`'s help page (`?library`):

> If library is called with no package or help argument, it lists all available packages in the libraries specified by lib.loc, and returns the corresponding information in an object of class "libraryIQR". (The structure of this class may change in future versions.) Use .packages(all = TRUE) to obtain just the names of all available packages, and installed.packages() for even more information.

Because the `package` and `help` argument from `library()` do not show a default value, it's easy to overlook the possibility to call `library()` without these arguments. (Instead of providing `NULL`s as default values `library()` uses `missing()` to check if these arguments were provided.)

```
str(formals(library))
#> Dotted pair list of 13
#>  $ package         : symbol
#>  $ help            : symbol
#>  $ pos             : num 2
#>  $ lib.loc         : NULL
#>  $ character.only  : logi FALSE
#>  $ logical.return  : logi FALSE
#>  $ warn.conflicts  : symbol
#>  $ quietly         : logi FALSE
#>  $ verbose         : language getOption("verbose")
#>  $ mask.ok         : symbol
#>  $ exclude         : symbol
#>  $ include.only    : symbol
#>  $ attach.required: language missing(include.only)
```

6.6 ... (dot-dot-dot)

Q1: Explain the following results:

```
sum(1, 2, 3)
#> [1] 6
mean(1, 2, 3)
#> [1] 1

sum(1, 2, 3, na.omit = TRUE)
#> [1] 7
mean(1, 2, 3, na.omit = TRUE)
#> [1] 1
```

A: Let's inspect the arguments and their order for both functions. For `sum()` these are ... and `na.rm`:

```
str(sum)
#> function (..., na.rm = FALSE)
```

For the ... argument `sum()` expects numeric, complex, or logical vector input (see `?sum`). Unfortunately, when ... is used, misspelled arguments (!) like `na.omit` won't raise an error (in case of no further input checks). So instead, na.omit is treated as a logical and becomes part of the ... argument. It will be coerced to 1 and be part of the sum. All other arguments are left unchanged. Therefore sum(1, 2, 3) returns 6 and sum(1, 2, 3, na.omit = TRUE) returns 7.

In contrast, the generic function mean() expects x, trim, na.rm and ... for its default method.

```
str(mean.default)
#> function (x, trim = 0, na.rm = FALSE, ...)
```

As na.omit is not one of mean()'s named arguments (x; and no candidate for partial matching), na.omit again becomes part of the ... argument. However, in contrast to sum() the elements of ... are not "part" of the mean. The other supplied arguments are matched by their order, i.e. x = 1, trim = 2 and na.rm = 3. As x is of length 1 and not NA, the settings of trim and na.rm do not affect the calculation of the mean. Both calls (mean(1, 2, 3) and mean(1, 2, 3, na.omit = TRUE)) return 1.

Q2: Explain how to find the documentation for the named arguments in the following function call:

```
plot(1:10, col = "red", pch = 20, xlab = "x", col.lab = "blue")
```

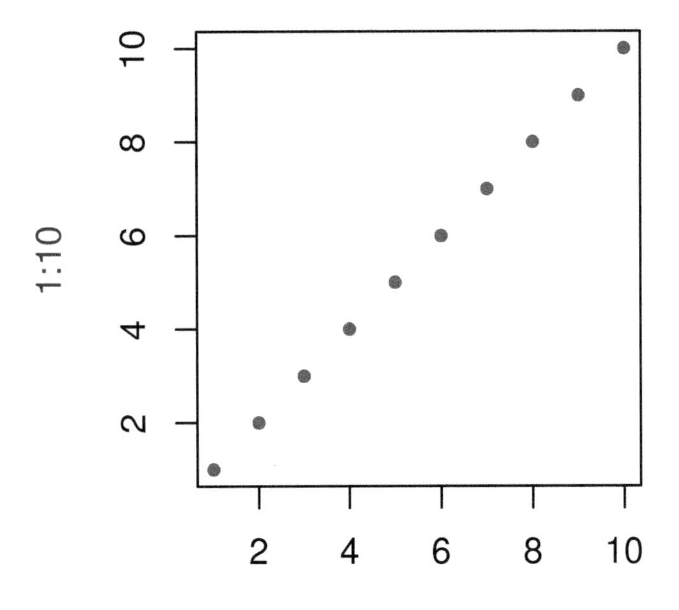

A: First we type `?plot` in the console and check the "Usage" section which contains:

```
plot(x, y, ...)
```

The arguments we want to learn more about (`col`, `pch`, `xlab`, `col.lab`) are part of the ... argument. There we can find information for the `xlab` argument and a recommendation to visit `?par` for the other arguments. Under `?par` we type "col" into the search bar, which leads us to the section "Color Specification". We also search for the `pch` argument, which leads to the recommendation to check `?points`. Finally, `col.lab` is also directly documented within `?par`.

Q3: Why does `plot(1:10, col = "red")` only colour the points, not the axes or labels? Read the source code of `plot.default()` to find out.

A: To learn about the internals of `plot.default()` we add `browser()` to the first line of the code and interactively run `plot(1:10, col = "red")`. This way we can see how the plot is built and learn where the axes are added.

This leads us to the function call

```
localTitle(main = main, sub = sub, xlab = xlab, ylab = ylab, ...)
```

The `localTitle()` function was defined in the first lines of `plot.default()` as:

```
localTitle <- function(..., col, bg, pch, cex, lty, lwd) title(...)
```

The call to `localTitle()` passes the `col` parameter as part of the ... argument to `title()`. `?title` tells us that the `title()` function specifies four parts of the plot: Main (title of the plot), sub (sub-title of the plot) and both axis labels. Therefore, it would introduce ambiguity inside `title()` to use `col` directly. Instead, one has the option to supply `col` via the ... argument, via `col.lab` or as part of `xlab` in the form `xlab = list(c("index"), col = "red")` (similar for `ylab`).

6.7 Exiting a function

Q1: What does `load()` return? Why don't you normally see these values?

A: `load()` loads objects saved to disk in `.Rdata` files by `save()`. When run successfully, `load()` invisibly returns a character vector containing the names of the newly loaded objects. To print these names to the console, one can set the argument `verbose` to `TRUE` or surround the call in parentheses to trigger R's auto-printing mechanism.

Q2: What does `write.table()` return? What would be more useful?

A: `write.table()` writes an object, usually a data frame or a matrix, to disk. The function invisibly returns `NULL`. It would be more useful if `write.table()` would (invisibly) return the input data, `x`. This would allow to save intermediate results and directly take on further processing steps without breaking the flow of the code (i.e. breaking it into different lines). One package which uses this pattern is the `{readr}` package [Wickham and Hester, 2020], which is part of the tidyverse-ecosystem (`https://www.tidyverse.org/`).

Q3: How does the `chdir` parameter of `source()` compare to `with_dir()`? Why might you prefer one to the other?

A: The `with_dir()` approach was given in *Advanced R* as:

```
with_dir <- function(dir, code) {
  old <- setwd(dir)
  on.exit(setwd(old))

  force(code)
}
```

`with_dir()` takes a path for a working directory (`dir`) as its first argument. This is the directory where the provided code (`code`) should be executed. Therefore, the current working directory is changed in `with_dir()` via `setwd()`. Then, `on.exit()` ensures that the modification of the working directory is reset to the initial value when the function exits. By passing the path explicitly, the user has full control over the directory to execute the code in.

In `source()` the code is passed via the `file` argument (a path to a file). The `chdir` argument specifies if the working directory should be changed to the directory containing the file. The default for `chdir` is `FALSE`, so you don't have to provide a value. However, as you can only provide `TRUE` or `FALSE`, you are also less flexible in choosing the working directory for the code execution.

Q4: Write a function that opens a graphics device, runs the supplied code, and closes the graphics device (always, regardless of whether or not the plotting code works).

A: To control the graphics device we use `pdf()` and `dev.off()`. To ensure a clean termination `on.exit()` is used.

```
plot_pdf <- function(code) {
  pdf("test.pdf")
```

```
  on.exit(dev.off(), add = TRUE)
  code
}
```

Q5: We can use `on.exit()` to implement a simple version of `capture.output()`.

```
capture.output2 <- function(code) {
  temp <- tempfile()
  on.exit(file.remove(temp), add = TRUE, after = TRUE)

  sink(temp)
  on.exit(sink(), add = TRUE, after = TRUE)

  force(code)
  readLines(temp)
}
capture.output2(cat("a", "b", "c", sep = "\n"))
#> [1] "a" "b" "c"
```

Compare `capture.output()` to `capture.output2()`. How do the functions differ? What features have I removed to make the key ideas easier to see? How have I rewritten the key ideas to be easier to understand?

A: Using `body(capture.output)` we inspect the source code of the original `capture.output()` function: The implementation for `capture.output()` is quite a bit longer (39 lines vs. 7 lines).

In `capture_output2()` the code is simply forced, and the output is caught via `sink()` in a temporary file. An additional feature of `capture_output()` is that one can also capture messages by setting `type = "message"`. As this is internally forwarded to `sink()`, this behaviour (and also `sink()`'s `split` argument) could be easily introduced within `capture_output2()` as well.

The main difference is that `capture.output()` calls print, i.e. compare the output of these two calls:

```
capture.output({1})
#> [1] "[1] 1"
capture.output2({1})
#> character(0)
```

6.8 Function forms

Q1: Rewrite the following code snippets into prefix form:

```
1 + 2 + 3
```

```
1 + (2 + 3)
```

```
if (length(x) <= 5) x[[5]] else x[[n]]
```

A: Let's rewrite the expressions to match the exact syntax from the code above. Because prefix functions already define the execution order, we may omit the parentheses in the second expression.

```
`+`(`+`(1, 2), 3)
```

```
`+`(1, `(`(`+`(2, 3)))
`+`(1, `+`(2, 3))
```

```
`if`(`<=`(length(x), 5), `[[`(x, 5), `[[`(x, n))
```

Q2: Clarify the following list of odd function calls:

```
x <- sample(replace = TRUE, 20, x = c(1:10, NA))
y <- runif(min = 0, max = 1, 20)
cor(m = "k", y = y, u = "p", x = x)
```

A: None of these functions provides a ... argument. Therefore, the function arguments are first matched exactly, then via partial matching and finally by position. This leads us to the following explicit function calls:

```
x <- sample(c(1:10, NA), size = 20, replace = TRUE)
y <- runif(20, min = 0, max = 1)
cor(x, y, use = "pairwise.complete.obs", method = "kendall")
```

Q3: Explain why the following code fails:

```
modify(get("x"), 1) <- 10
#> Error: target of assignment expands to non-language object
```

A: First, let's define `x` and recall the definition of `modify()` from *Advanced R*:

```
x <- 1:3

`modify<-` <- function(x, position, value) {
  x[position] <- value
  x
}
```

R internally transforms the code, and the transformed code reproduces the error above:

```
get("x") <- `modify<-`(get("x"), 1, 10)
#> Error in get("x") <- `modify<-`(get("x"), 1, 10) :
#>   target of assignment expands to non-language object
```

The error occurs during the assignment because no corresponding replacement function, i.e. `get<-`, exists for `get()`. To confirm this, we reproduce the error via the following simplified example.

```
get("x") <- 2
#> Error in get("x") <- 2 :
#>   target of assignment expands to non-language object
```

Q4: Create a replacement function that modifies a random location in a vector.

A: Let's define `random<-` like this:

```
`random<-` <- function(x, value) {
  idx <- sample(length(x), 1)
  x[idx] <- value
  x
}
```

Q5: Write your own version of `+` that pastes its inputs together if they are character vectors but behaves as usual otherwise. In other words, make this code work:

```
1 + 2
#> [1] 3

"a" + "b"
#> [1] "ab"
```

A: To achieve this behaviour, we need to override the `+` operator. We need to take care to not use the `+` operator itself inside of the function definition, as this would lead to an undesired infinite recursion. We also add `b = 0L` as a default value to keep the behaviour of `+` as a unary operator, i.e. to keep `+ 1` working and not throwing an error.

```
`+` <- function(a, b = 0L) {
  if (is.character(a) && is.character(b)) {
    paste0(a, b)
  } else {
    base::`+`(a, b)
  }
}

# Test
+ 1
#> [1] 1
1 + 2
#> [1] 3
"a" + "b"
#> [1] "ab"

# Return back to the original `+` operator
rm(`+`)
```

Q6: Create a list of all the replacement functions found in the `{base}` package. Which ones are primitive functions? (Hint use `apropos()`)

A: The hint suggests to look for functions with a specific naming pattern: Replacement functions conventionally end on "<-". We can search for these objects by supplying the regular expression `"<-$"` to `apropos()`. `apropos()` also allows to return the position on the search path (`search()`) for each of its matches via setting `where = TRUE`. Finally, we can set `mode = function` to narrow down our search to relevant objects only. This gives us the following statement to begin with:

```
repls <- apropos("<-", where = TRUE, mode = "function")
head(repls, 30)
#>                    10                     10                   10
#>        ".rowNamesDF<-"                 "[[<-"   "[[<-.data.frame"
#>                    10                     10                   10
#>      "[[<-.factor" "[[<-.numeric_version"       "[[<-.POSIXlt"
#>                    10                     10                   10
#>                 "[<-"       "[<-.data.frame"          "[<-.Date"
```

```
#>                  10                       10                      10
#>       "[<-.factor" "[<-.numeric_version"           "[<-.POSIXct"
#>                  10                       10                      10
#>      "[<-.POSIXlt"                   "@<-"                   "<-"
#>                  10                       10                      10
#>              "<<-"                   "$<-"      "$<-.data.frame"
#>                   8                       10                      10
#>             "as<-"                "attr<-"        "attributes<-"
#>                   8                       10                      10
#>           "body<-"                "body<-"             "class<-"
#>                   8                       10                      10
#>         "coerce<-"            "colnames<-"           "comment<-"
#>                   3                       10                      10
#>      "contrasts<-"                "diag<-"               "dim<-"
```

To restrict repl to names of replacement functions from the {base} package, we select only matches containing the relevant position on the search path.

```
repls_base <- repls[names(repls) == length(search())]
repls_base
#>                    10                       10
#>        ".rowNamesDF<-"                   "[[<-"
#>                    10                       10
#>     "[[<-.data.frame"          "[[<-.factor"
#>                    10                       10
#> "[[<-.numeric_version"          "[[<-.POSIXlt"
#>                    10                       10
#>                 "[<-"      "[<-.data.frame"
#>                    10                       10
#>           "[<-.Date"           "[<-.factor"
#>                    10                       10
#>  "[<-.numeric_version"          "[<-.POSIXct"
#>                    10                       10
#>        "[<-.POSIXlt"                  "@<-"
#>                    10                       10
#>                  "<-"                  "<<-"
#>                    10                       10
#>                 "$<-"      "$<-.data.frame"
#>                    10                       10
#>              "attr<-"        "attributes<-"
#>                    10                       10
#>              "body<-"             "class<-"
#>                    10                       10
```

```
#>             "colnames<-"                "comment<-"
#>                       10                         10
#>                 "diag<-"                    "dim<-"
#>                       10                         10
#>             "dimnames<-"  "dimnames<-.data.frame"
#>                       10                         10
#>             "Encoding<-"            "environment<-"
#>                       10                         10
#>              "formals<-"                  "is.na<-"
#>                       10                         10
#>        "is.na<-.default"         "is.na<-.factor"
#>                       10                         10
#> "is.na<-.numeric_version"              "length<-"
#>                       10                         10
#>         "length<-.Date"        "length<-.difftime"
#>                       10                         10
#>       "length<-.factor"         "length<-.POSIXct"
#>                       10                         10
#>      "length<-.POSIXlt"                 "levels<-"
#>                       10                         10
#>       "levels<-.factor"                   "mode<-"
#>                       10                         10
#>      "mostattributes<-"                  "names<-"
#>                       10                         10
#>       "names<-.POSIXlt"               "oldClass<-"
#>                       10                         10
#>          "parent.env<-"             "regmatches<-"
#>                       10                         10
#>           "row.names<-"  "row.names<-.data.frame"
#>                       10                         10
#>    "row.names<-.default"              "rownames<-"
#>                       10                         10
#>               "split<-"      "split<-.data.frame"
#>                       10                         10
#>       "split<-.default"          "storage.mode<-"
#>                       10                         10
#>              "substr<-"             "substring<-"
#>                       10                         10
#>               "units<-"        "units<-.difftime"
```

To find out which of these functions are primitives, we first search for these functions via `mget()` and then subset the result using `Filter()` and `is.primitive()`.

```
repls_base_prim <- mget(repls_base, envir = baseenv()) %>%
  Filter(is.primitive, .) %>%
  names()

repls_base_prim
#>  [1] "[[<-"           "[<-"             "@<-"
#>  [4] "<-"             "<<-"             "$<-"
#>  [7] "attr<-"         "attributes<-"    "class<-"
#> [10] "dim<-"          "dimnames<-"      "environment<-"
#> [13] "length<-"       "levels<-"        "names<-"
#> [16] "oldClass<-"     "storage.mode<-"
```

Overall the {base} package contains 62 replacement functions of which 17 are primitive functions.

Q7: What are valid names for user-created infix functions?

A: Let's cite from the section on function forms (https://adv-r.hadley.nz/functions.html#function-forms) from *Advanced R*:

> … names of infix functions are more flexible than regular R functions: they can contain any sequence of characters except "%".

Q8: Create an infix `xor()` operator.

A: We could create an infix `%xor%` like this:

```
`%xor%` <- function(a, b) {
  xor(a, b)
}
TRUE %xor% TRUE
#> [1] FALSE
FALSE %xor% TRUE
#> [1] TRUE
```

Q9: Create infix versions of the set functions `intersect()`, `union()`, and `setdiff()`. You might call them `%n%`, `%u%`, and `%/%` to match conventions from mathematics.

A: These infix operators could be defined in the following way. (`%/%` is chosen instead of `%\%`, because `\` serves as an escape character.)

```
`%n%` <- function(a, b) {
  intersect(a, b)
}
```

```
`%u%` <- function(a, b) {
  union(a, b)
}

`%/%` <- function(a, b) {
  setdiff(a, b)
}

x <- c("a", "b", "d")
y <- c("a", "c", "d")

x %u% y
#> [1] "a" "b" "d" "c"
x %n% y
#> [1] "a" "d"
x %/% y
#> [1] "b"
```

7

Environments

Prerequisites

Just like in *Advanced R*, we mainly use the `{rlang}` package [Henry and Wickham, 2020b] to work with environments.

```
library(rlang)
```

7.2 Environment basics

Q1: List three ways in which an environment differs from a list.

A: The most important differences between environments and lists are:

- environments have reference semantics (i.e. they don't copy-on-modify)
- environments have parents
- the contents of an environment must have unique names
- the contents of an environment are not ordered
- (environments can only be compared via `identical()`; not with `==`)
- (environments can contain themselves)

Q2: Create an environment as illustrated by this picture.

DOI: 10.1201/9781003175414-7

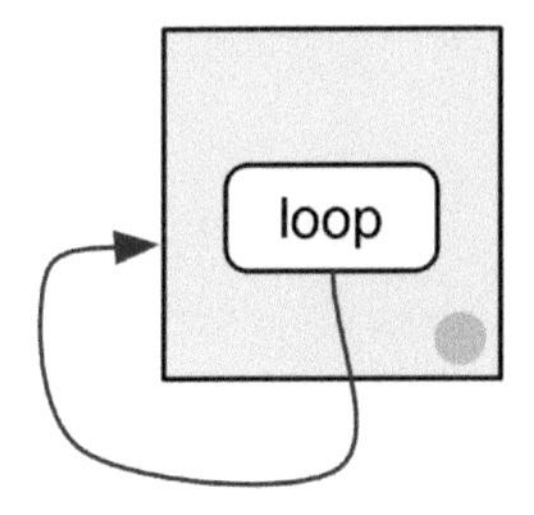

A: Let's create an environment that contains itself.

```
e1 <- env()
e1$loop <- e1

# Print the environment
env_print(e1)
#> <environment: 0x55829c4822d0>
#> parent: <environment: global>
#> bindings:
#>  * loop: <env>

# Verify that it contains itself
lobstr::ref(e1)
#> █ [1:0x55829c4822d0] <env>
#> └─loop = [1:0x55829c4822d0]
```

Q3: Create a pair of environments as illustrated by this picture.

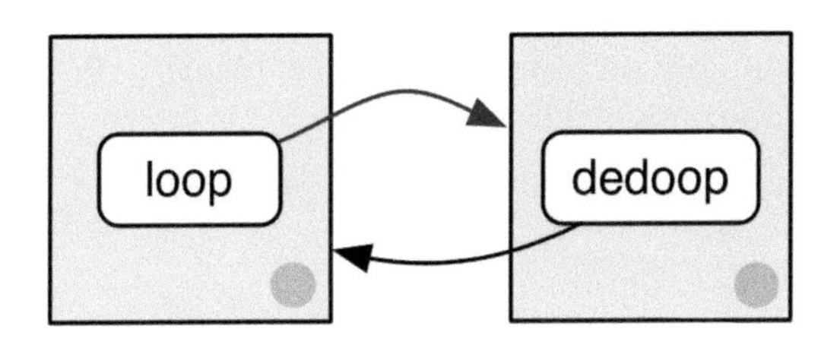

A: These two environments contain each other:

```
e1 <- env()
e2 <- env()

e1$loop   <- e2
```

```
e2$dedoop <- e1

lobstr::ref(e1)
#> █ [1:0x55829be85fc8] <env>
#> └─loop = █ [2:0x55829b601b78] <env>
#>          └─dedoop = [1:0x55829be85fc8]
lobstr::ref(e2)
#> █ [1:0x55829b601b78] <env>
#> └─dedoop = █ [2:0x55829be85fc8] <env>
#>            └─loop = [1:0x55829b601b78]
```

Q4: Explain why `e[[1]]` and `e[c("a", "b")]` don't make sense when `e` is an environment.

A: The first option doesn't make sense, because elements of an environment are not ordered. The second option would return two objects at the same time. What data structure would they be contained inside?

Q5: Create a version of `env_poke()` that will only bind new names, never re-bind old names. Some programming languages only do this, and are known as single assignment languages (`http://en.wikipedia.org/wiki/Assignment_(computer_science)#Single_assignment`).

A: As described in *Advanced R* `rlang::env_poke()` takes a name (as string) and a value to assign (or reassign) a binding in an environment.

```
e3 <- new.env()

env_poke(e3, "a", 100)
e3$a
#> [1] 100
env_poke(e3, "a", 200)
e3$a
#> [1] 200
```

So, we want `env_poke2()` to test, if the supplied name is already present in the given environment. This can be checked via `env_has()`. If this is the case, an (informative) error is thrown.

```
env_poke2 <- function(env, name, value) {
  if (env_has(env, name)) {
    abort(paste0("\"", name, "\" is already assigned to a value."))
  }
```

```
  env_poke(env, name, value)
  invisible(env)
}

# Test
env_poke2(e3, "b", 100)
e3$b
#> [1] 100
env_poke2(e3, "b", 200)
#> Error: "b" is already assigned to a value.
```

Q6: What does this function do? How does it differ from `<<-` and why might you prefer it?

```
rebind <- function(name, value, env = caller_env()) {
  if (identical(env, empty_env())) {
    stop("Can't find `", name, "`", call. = FALSE)
  } else if (env_has(env, name)) {
    env_poke(env, name, value)
  } else {
    rebind(name, value, env_parent(env))
  }
}
rebind("a", 10)
#> Error: Can't find `a`
a <- 5
rebind("a", 10)
a
#> [1] 10
```

A: The primary difference between `rebind()` and `<<-` is that `rebind()` will only carry out an assignment when it finds an existing binding; unlike `<<-` it will never create a new one in the global environment. This behaviour of `<<-` is usually undesirable because global variables introduce non-obvious dependencies between functions.

7.3 Recursing over environments

Q1: Modify `where()` to return *all* environments that contain a binding for `name`. Carefully think through what type of object the function will need to return.

A: `where()` searches (recursively) for a given name within a given environment and its ancestors. If `where()` finds the name in one of these environments, it returns the environment's name. Otherwise, it throws an error.

The definition of `where()` was given in *Advanced R* as:

```
where <- function(name, env = caller_env()) {
  if (identical(env, empty_env())) {
    # Base case
    stop("Can't find `", name, "`.", call. = FALSE)
  } else if (env_has(env, name)) {
    # Success case
    env
  } else {
    # Recursive case
    where(name, env_parent(env))
  }
}
```

Our modified version of `where()` will always recurse until it reaches the empty environment. No matter if it has already found the name or not. Along the way, it will check each environment for the given `name`. Finally, it will return a list of environments where the binding was found; if no binding was found, the list will be empty.

Please also note how the list is initialised via the default argument, when the function is called for the first time. This is a bit confusing, which is why it's common to wrap a recursive function inside another, more user friendly, function.

```
where2 <- function(name, env = caller_env(), results = list()) {
  if (identical(env, empty_env())) {
    # Base case
    results
  } else {
    # Recursive case
    if (env_has(env, name)) {
```

```
      results <- c(results, env)
    }
    where2(name, env_parent(env), results)
  }
}

# Test
e1a <- env(empty_env(), a = 1, b = 2)
e1b <- env(e1a, b = 10, c = 11)
e1c <- env(e1b, a = 12, d = 13)

where2("a", e1c)
#> [[1]]
#> <environment: 0x55829f430ca0>
#>
#> [[2]]
#> <environment: 0x55829f327aa0>
```

Q2: Write a function called `fget()` that finds only function objects. It should have two arguments, `name` and `env`, and should obey the regular scoping rules for functions: if there's an object with a matching name that's not a function, look in the parent. For an added challenge, also add an `inherits` argument which controls whether the function recurses up the parents or only looks in one environment.

A: We follow a similar approach to the previous exercise. This time we additionally check if the found object is a function and implement an argument to turn off the recursion, if desired.

```
fget <- function(name, env = caller_env(), inherits = TRUE) {
  # Base case
  if (env_has(env, name)) {
    obj <- env_get(env, name)

    if (is.function(obj)) {
      return(obj)
    }
  }

  if (identical(env, emptyenv()) || !inherits) {
    stop("Could not find a function called \"", name, "\".",
      call. = FALSE
    )
  }
```

```
  # Recursive Case
  fget(name, env_parent(env))
}

# Test
mean <- 10
fget("mean", inherits = TRUE)
#> function (x, ...)
#> UseMethod("mean")
#> <bytecode: 0x55829be34488>
#> <environment: namespace:base>
```

7.4 Special environments

Q1: How is `search_envs()` different to `env_parents(global_env())`?

A: `search_envs()` returns all the environments on the search path, which is "a chain of environments containing exported functions of attached packages" (from `?search_envs`). Every time you attach a new package, this search path will grow. The search path ends with the base-environment. The global environment is included, because functions present in the global environment will always be part of the search path.

```
search_envs()
#>  [[1]] $ <env: global>
#>  [[2]] $ <env: package:rlang>
#>  [[3]] $ <env: package:stats>
#>  [[4]] $ <env: package:graphics>
#>  [[5]] $ <env: package:grDevices>
#>  [[6]] $ <env: package:utils>
#>  [[7]] $ <env: package:datasets>
#>  [[8]] $ <env: package:methods>
#>  [[9]] $ <env: Autoloads>
#> [[10]] $ <env: package:base>
```

`env_parents(global_env())` will list all the ancestors of the global environment, therefore the global environment itself is not included. This also includes the "ultimate ancestor", the empty environment. This environment is not considered part of the search path because it contains no objects.

```
env_parents(global_env())
#>  [[1]] $ <env: package:rlang>
#>  [[2]] $ <env: package:stats>
#>  [[3]] $ <env: package:graphics>
#>  [[4]] $ <env: package:grDevices>
#>  [[5]] $ <env: package:utils>
#>  [[6]] $ <env: package:datasets>
#>  [[7]] $ <env: package:methods>
#>  [[8]] $ <env: Autoloads>
#>  [[9]] $ <env: package:base>
#> [[10]] $ <env: empty>
```

Q2: Draw a diagram that shows the enclosing environments of this function:

```
f1 <- function(x1) {
  f2 <- function(x2) {
    f3 <- function(x3) {
      x1 + x2 + x3
    }
    f3(3)
  }
  f2(2)
}
f1(1)
```

A: This exercise urges us to think carefully about the function environment at creation time.

When `f1` is defined it binds its parent environment, which is the global environment. But `f2` will only be created at runtime of `f1` and will therefore bind `f1`'s execution environment. The value `1` will also bind to the name `x1` at execution time. The same holds true for `x2`, `f3` and `x3`.

The following diagram visualises the relations between the function environments.

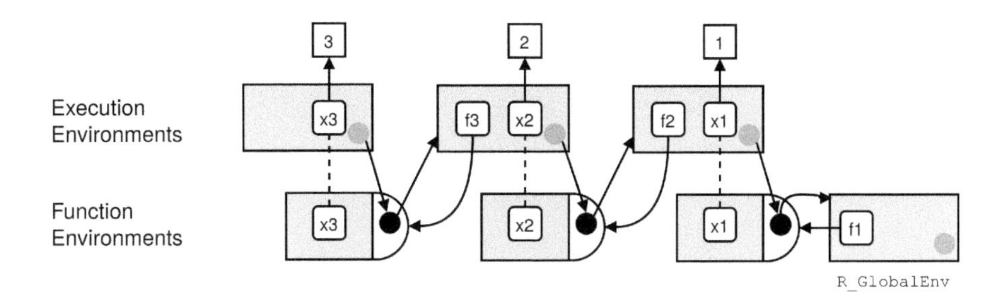

We can also inspect the binding of the environments, adding print statements to the function definition. Please note that these print statements will be evaluated at execution time. Therefore, the execution of `f1(1)` will print different results each time we run it.

```
f1 <- function(x1) {
  f2 <- function(x2) {
    f3 <- function(x3) {
      x1 + x2 + x3
      print("f3")
      print(env_print())
    }
    f3(3)
    print("f2")
    print(env_print())
  }
  f2(2)
  print("f1")
  print(env_print())
}

f1(1)
#> [1] "f3"
#> <environment: 0x55829e3a0eb0>
#> parent: <environment: 0x55829e3a1070>
#> bindings:
#>  * x3: <dbl>
#> <environment: 0x55829e3a0eb0>
#> [1] "f2"
#> <environment: 0x55829e3a1070>
#> parent: <environment: 0x55829e3a1230>
#> bindings:
#>  * f3: <fn>
```

```
#>   * x2: <dbl>
#> <environment: 0x55829e3a1070>
#> [1] "f1"
#> <environment: 0x55829e3a1230>
#> parent: <environment: global>
#> bindings:
#>   * f2: <fn>
#>   * x1: <dbl>
#> <environment: 0x55829e3a1230>
```

Q3: Write an enhanced version of `str()` that provides more information about functions. Show where the function was found and what environment it was defined in.

A: To solve this problem, we need to write a function that takes the name of a function and looks for that function returning both the function and the environment that it was found in.

```
fget2 <- function(name, env = caller_env()) {
  # Base case
  if (env_has(env, name)) {
    obj <- env_get(env, name)

    if (is.function(obj)) {
      return(list(fun = obj, env = env))
    }
  }

  if (identical(env, emptyenv())) {
    stop("Could not find a function called \"", name, "\"",
      call. = FALSE
    )
  }

  # Recursive Case
  fget2(name, env_parent(env))
}

fstr <- function(fun_name, env = caller_env()) {
  if (!is.character(fun_name) && length(fun_name) == 1) {
    stop("`fun_name` must be a string.", call. = FALSE)
  }
  fun_env <- fget2(fun_name, env)
```

```
  list(
    where = fun_env$env,
    enclosing = fn_env(fun_env$fun)
  )
}

# Test
fstr("mean")
#> $where
#> <environment: base>
#>
#> $enclosing
#> <environment: namespace:base>
```

Once you have learned about tidy evaluation, you could rewrite `fstr()` to use `enquo()` so that you'd call it more like `str()`, i.e. `fstr(sum)`.

7.5 Call stacks

Q1: Write a function that lists all the variables defined in the environment in which it was called. It should return the same results as `ls()`.

A: We can implement this dynamic scoping behaviour by explicitly referencing the caller environment. Please note that this approach returns also variables starting with a dot, an option that `ls()` usually requires.

```
ls2 <- function(env = caller_env()) {
  sort(env_names(env))
}

# Test in global environment
ls(all.names = TRUE)
#>  [1] ".Random.seed" "%>%"          "a"            "e1"
#>  [5] "e1a"          "e1b"          "e1c"          "e2"
#>  [9] "e3"           "env_poke2"    "error_wrap"   "f1"
#> [13] "fget"         "fget2"        "fstr"         "ls2"
#> [17] "mean"         "rebind"       "where"        "where2"
ls2()
#>  [1] ".Random.seed" "%>%"          "a"            "e1"
#>  [5] "e1a"          "e1b"          "e1c"          "e2"
```

```
#>  [9] "e3"          "env_poke2"    "error_wrap"   "f1"
#> [13] "fget"        "fget2"        "fstr"         "ls2"
#> [17] "mean"        "rebind"       "where"        "where2"

# Test in "sandbox" environment
e1 <- env(a = 1, b = 2)
ls(e1)
#> [1] "a" "b"
ls2(e1)
#> [1] "a" "b"
```

8

Conditions

Prerequisites

Similar to the environments chapter, we also use functions from the `{rlang}` package to work with conditions.

```
library(rlang)
```

8.2 Signalling conditions

Q1: Write a wrapper around `file.remove()` that throws an error if the file to be deleted does not exist.

A: We prefer the following solution for its clarity and simplicity:

```
file_remove_strict <- function(path) {
  if (!file.exists(path)) {
    stop("Can't delete the file \"", path,
         "\" because it doesn't exist.",
         call. = FALSE
    )
  }
  file.remove(path)
}

# Test
saveRDS(mtcars, "mtcars.rds")
file_remove_strict("mtcars.rds")
#> [1] TRUE
```

DOI: 10.1201/9781003175414-8

```r
file_remove_strict("mtcars.rds")
#> Error: Can't delete the file "mtcars.rds" because it doesn't exist.
```

Q2: What does the `appendLF` argument to `message()` do? How is it related to `cat()`?

A: The `appendLF` argument automatically appends a new line to the message. Let's illustrate this behaviour with a small example function:

```r
multiline_msg <- function(appendLF = TRUE) {
  message("first", appendLF = appendLF)
  message("second", appendLF = appendLF)
  cat("third")
  cat("fourth")
}

multiline_msg(appendLF = TRUE)
#> first
#> second
#> thirdfourth
multiline_msg(appendLF = FALSE)
#> firstsecondthirdfourth
```

Comparable behaviour regarding line breaks for `cat()` can be achieved via setting its `sep` argument to `"\n"`.

8.4 Handling conditions

Q1: What extra information does the condition generated by `abort()` contain compared to the condition generated by `stop()`, i.e. what's the difference between these two objects? Read the help for `?abort` to learn more.

```r
catch_cnd(stop("An error"))
catch_cnd(abort("An error"))
```

A: In contrast to `stop()`, which contains the call, `abort()` stores the whole backtrace generated by `rlang::trace_back()`. This is a lot of extra data!

```
str(catch_cnd(stop("An error")))
#> List of 2
#>  $ message: chr "An error"
#>  $ call   : language force(expr)
#>  - attr(*, "class")= chr [1:3] "simpleError" "error" "condition"

str(catch_cnd(abort("An error")))
#> List of 3
#>  $ message: chr "An error"
#>  $ trace  :List of 3
#>   ..$ calls  :List of 8
#>   .. ..$ : language utils::str(catch_cnd(abort("An error")))
#>   .. ..$ : language rlang::catch_cnd(abort("An error"))
#>   .. ..$ : language rlang::eval_bare(rlang::expr(tryCatch(!!!handle..
#>   .. ..$ : language base::tryCatch(condition = function (x)  x, { ...
#>   .. ..$ : language base:::tryCatchList(expr, classes, parentenv, h..
#>   .. ..$ : language base:::tryCatchOne(expr, names, parentenv, hand..
#>   .. ..$ : language base:::doTryCatch(return(expr), name, parentenv..
#>   .. ..$ : language base::force(expr)
#>   ..$ parents: int [1:8] 0 0 2 2 4 5 6 2
#>   ..$ indices: int [1:8] 26 27 28 29 30 31 32 33
#>   ..- attr(*, "class")= chr "rlang_trace"
#>   ..- attr(*, "version")= int 1
#>  $ parent : NULL
#>  - attr(*, "class")= chr [1:3] "rlang_error" "error" "condition"
```

Q2: Predict the results of evaluating the following code

```
show_condition <- function(code) {
  tryCatch(
    error = function(cnd) "error",
    warning = function(cnd) "warning",
    message = function(cnd) "message",
    {
      code
      NULL
    }
  )
}
```

```
show_condition(stop("!"))
show_condition(10)
```

```r
show_condition(warning("?!"))
show_condition({
  10
  message("?")
  warning("?!")
})
```

A: The first three examples are straightforward:

```r
show_condition(stop("!"))       # stop raises an error
#> [1] "error"
show_condition(10)              # no condition is signalled
#> NULL
show_condition(warning("?!"))  # warning raises a warning
#> [1] "warning"
```

The last example is the most interesting and makes us aware of the exiting qualities of `tryCatch()`; it will terminate the evaluation of the code as soon as it is called.

```r
show_condition({
  10
  message("?")
  warning("?!")
})
#> [1] "message"
```

Q3: Explain the results of running this code:

```r
withCallingHandlers(  # (1)
  message = function(cnd) message("b"),
  withCallingHandlers(  # (2)
    message = function(cnd) message("a"),
    message("c")
  )
)
#> b
#> a
#> b
#> c
```

A: It's a little tricky to untangle the flow here:

First, `message("c")` is run, and it's caught by (1). It then calls `message("a")`, which is caught by (2), which calls `message("b")`. `message("b")` isn't caught by anything, so we see a `b` on the console, followed by `a`. But why do we get another `b` before we see `c`? That's because we haven't handled the message, so it bubbles up to the outer calling handler.

Q4: Read the source code for `catch_cnd()` and explain how it works. At the time *Advanced R* was written, the source for `catch_cnd()` was a little simpler:

```
catch_cnd <- function(expr) {
  tryCatch(
    condition = function(cnd) cnd,
    {
      force(expr)
      return(NULL)
    }
  )
}
```

A: `catch_cnd()` is a simple wrapper around `tryCatch()`. If a condition is signalled, it's caught and returned. If no condition is signalled, execution proceeds sequentially and the function returns `NULL`.

The current version of `catch_cnd()` is a little more complex because it allows you to specify which classes of condition you want to capture. This requires some manual code generation because the interface of `tryCatch()` provides condition classes as argument names.

```
rlang::catch_cnd
#> function (expr, classes = "condition")
#> {
#>     stopifnot(is_character(classes))
#>     handlers <- rep_named(classes, list(identity))
#>     eval_bare(rlang::expr(tryCatch(!!!handlers, {
#>         force(expr)
#>         return(NULL)
#>     })))
#> }
#> <bytecode: 0x557856fce1c0>
#> <environment: namespace:rlang>
```

Q5: How could you rewrite `show_condition()` to use a single handler?

A: `show_condition()` was defined in one of the previous questions. Let's use the `condition` argument of `tryCatch()` as shown in `rlang::catch_cond()` above for our re-implementation:

```
show_condition2 <- function(code) {
  tryCatch(
    condition = function(cnd) {
      if (inherits(cnd, "error"))   return("error")
      if (inherits(cnd, "warning")) return("warning")
      if (inherits(cnd, "message")) return("message")
    },
    {
      code
      NULL
    }
  )
}

# Test
show_condition2(stop("!"))
#> [1] "error"
show_condition2(10)
#> NULL
show_condition2(warning("?!"))
#> [1] "warning"
show_condition2({
  10
  message("?")
  warning("?!")
})
#> [1] "message"
```

`tryCatch()` executes the code and captures any condition raised. The function provided as the `condition` handles this condition. In this case it dispatches on the class of the condition.

8.5 Custom conditions

Q1: Inside a package, it's occasionally useful to check that a package is installed before using it. Write a function that checks if a package is installed (with `requireNamespace("pkg", quietly = FALSE))` and if not, throws a custom condition that includes the package name in the metadata.

A: We use `rlang::abort()` to supply error metadata:

```r
check_installed <- function(package) {
  if (!requireNamespace(package, quietly = FALSE)) {
    abort(
      "error_pkg_not_found",
      message = paste0("package '", package, "' not installed."),
      package = package
    )
  }

  TRUE
}

check_installed("ggplot2")
#> [1] TRUE
check_installed("ggplot3")
#> Loading required namespace: ggplot3
#> Error: package 'ggplot3' not installed.
```

Q2: Inside a package you often need to stop with an error when something is not right. Other packages that depend on your package might be tempted to check these errors in their unit tests. How could you help these packages to avoid relying on the error message which is part of the user interface rather than the API and might change without notice?

A: Instead of returning an error it might be preferable to throw a customised condition and place a standardised error message inside the metadata. Then the downstream package could check for the class of the condition, rather than inspecting the message.

8.6 Applications

Q1: Create `suppressConditions()` that works like `suppressMessages()` and `suppressWarnings()` but suppresses everything. Think carefully about how you should handle errors.

A: In general, we would like to catch errors, since they contain important information for debugging. To suppress the error message and hide the returned error object from the console, we handle errors within a `tryCatch()` and return the error object invisibly:

```
suppressErrors <- function(expr) {
  tryCatch(
    error = function(cnd) invisible(cnd),
    interrupt = function(cnd) {
      stop("Terminated by the user.",
        call. = FALSE
      )
    },
    expr
  )
}
```

After we defined the error handling, we can just combine it with the other handlers to create `suppressConditions()`:

```
suppressConditions <- function(expr) {
  suppressErrors(suppressWarnings(suppressMessages(expr)))
}
```

To test the new function, we apply it to a set of conditions and inspect the returned error object.

```
# The messages/warnings/conditions are suppressed successfully
error_obj <- suppressConditions({
  message("message")
  warning("warning")
  abort("error")
})

error_obj
#> <error/rlang_error>
#> error
#> Backtrace:
#>    1. global::suppressConditions(...)
#>   12. base::suppressMessages(expr)
#>   13. base::withCallingHandlers(...)
```

Q2: Compare the following two implementations of `message2error()`. What is the main advantage of `withCallingHandlers()` in this scenario? (Hint: look carefully at the traceback.)

```r
message2error <- function(code) {
  withCallingHandlers(code, message = function(e) stop(e))
}
message2error <- function(code) {
  tryCatch(code, message = function(e) stop(e))
}
```

A: Both functions differ in the way conditions are handled. `withCallingHandlers()` creates a *calling handler*, which is executed from within the signalling function. This makes it possible to record a detailed call stack, which helps us identify the signalling condition.

`tryCatch()` defines an *exiting handler*, which means that the signalling function is terminated as soon as a condition is raised. It also returns control to the context where `tryCatch()` was called.

In this example the use of `withCallingHandlers()` returns more information than the use of `tryCatch()`. This allows us to determine the exact call that raised the condition.

```r
message2error1 <- function(code) {
  withCallingHandlers(code, message = function(e) stop("error"))
}

message2error1({1;  message("hidden error"); NULL})
#> Error in (function (e) : error
traceback()
#> 9: stop("error") at #2
#> 8: (function (e)
#>    stop("error"))(list(message = "hidden error\n",
#>      call = message("hidden error")))
#> 7: signalCondition(cond)
#> 6: doWithOneRestart(return(expr), restart)
#> 5: withOneRestart(expr, restarts[[1L]])
#> 4: withRestarts({
#>        signalCondition(cond)
#>        defaultHandler(cond)
#>    }, muffleMessage = function() NULL)
#> 3: message("hidden error") at #1
#> 2: withCallingHandlers(code,
#>      message = function(e) stop("error")) at #2
#> 1: message2error1({
#>        1
#>        message("hidden error")
```

```
#>         NULL
#>     })
```

```
message2error2 <- function(code) {
  tryCatch(code, message = function(e) (stop("error")))
}

message2error2({1; stop("hidden error"); NULL})
#> Error in value[[3L]](cond) : error
traceback()
#> 6: stop("error") at #2
#> 5: value[[3L]](cond)
#> 4: tryCatchOne(expr, names, parentenv, handlers[[1L]])
#> 3: tryCatchList(expr, classes, parentenv, handlers)
#> 2: tryCatch(code, message = function(e) (stop("error"))) at #2
#> 1: message2error2({
#>         1
#>         message("hidden error")
#>         NULL
#>     })
```

Q3: How would you modify the `catch_cnds()` definition if you wanted to recreate the original intermingling of warnings and messages?

A: It looks like Hadley wrote a part of the chapter after the exercises, as the `catch_cnds()` function defined in the chapter already solves this problem by storing all messages and warnings in their original order within a list.

```
catch_cnds <- function(expr) {
  conds <- list()
  add_cond <- function(cnd) {
    conds <<- append(conds, list(cnd))
    cnd_muffle(cnd)
  }

  tryCatch(
    error = function(cnd) {
      conds <<- append(conds, list(cnd))
    },
    withCallingHandlers(
      message = add_cond,
      warning = add_cond,
      expr
```

```
    )
  )

  conds
}

# Test
catch_cnds({
  inform("message a")
  warn("warning b")
  inform("message c")
})
#> [[1]]
#> <message: message a
#> >
#>
#> [[2]]
#> <warning: warning b>
#>
#> [[3]]
#> <message: message c
#> >
```

Q4: Why is catching interrupts dangerous? Run this code to find out.

```
bottles_of_beer <- function(i = 99) {
  message(
    "There are ", i,
    " bottles of beer on the wall, ", i,
    " bottles of beer."
  )
  while (i > 0) {
    tryCatch(
      Sys.sleep(1),
      interrupt = function(err) {
        i <<- i - 1
        if (i > 0) {
          message(
            "Take one down, pass it around, ", i,
            " bottle", if (i > 1) "s", " of beer on the wall."
          )
        }
      }
```

```
    )
  }
  message(
    "No more bottles of beer on the wall, ",
    "no more bottles of beer."
  )
}
```

A: When running the `bottles_of_beer()` function in your console, the output should look somehow like the following:

```
bottles_of_beer()
#> There are 99 bottles of beer on the wall, 99 bottles of beer.
#> Take one down, pass it around, 98 bottles of beer on the wall.
#> Take one down, pass it around, 97 bottles of beer on the wall.
#> Take one down, pass it around, 96 bottles of beer on the wall.
#> Take one down, pass it around, 95 bottles of beer on the wall.
#>
```

At this point you'll probably recognise how hard it is to get the number of bottles down from `99` to `0`. There's no way to break out of the function because we're capturing the interrupt that you'd usually use!

Part II

Functional programming

9

Functionals

Prerequisites

For the functional programming part of the book, we will mainly use functions from the `{purrr}` package [Henry and Wickham, 2020a].

```
library(purrr)
```

9.2 My first functional: `map()`

Q1: Use `as_mapper()` to explore how `{purrr}` generates anonymous functions for the integer, character, and list helpers. What helper allows you to extract attributes? Read the documentation to find out.

A: `map()` offers multiple ways (functions, formulas, and extractor functions) to specify its function argument (`.f`). Initially, the various inputs have to be transformed into a valid function, which is then applied. The creation of this valid function is the job of `as_mapper()` and it is called every time `map()` is used.

Given character, numeric or list input `as_mapper()` will create an extractor function. Characters select by name, while numeric input selects by positions and a list allows a mix of these two approaches. This extractor interface can be very useful, when working with nested data.

The extractor function is implemented as a call to `purrr::pluck()`, which accepts a list of accessors (accessors "access" some part of your data object).

```
as_mapper(c(1, 2))  # equivalent to function(x) x[[1]][[2]]
#> function (x, ...)
```

DOI: 10.1201/9781003175414-9

```
#> pluck(x, 1, 2, .default = NULL)
#> <environment: 0x55adc2679ca0>
as_mapper(c("a", "b")) # equivalent to function(x) x[["a"]][["b"]]
#> function (x, ...)
#> pluck(x, "a", "b", .default = NULL)
#> <environment: 0x55adc25c7ae0>
as_mapper(list(1, "b")) # equivalent to function(x) x[[1]][["b"]]
#> function (x, ...)
#> pluck(x, 1, "b", .default = NULL)
#> <environment: 0x55adc252e900>
```

Besides mixing positions and names, it is also possible to pass along an accessor function. This is basically an anonymous function that gets information about some aspect of the input data. You are free to define your own accessor functions.

If you need to access certain attributes, the helper `attr_getter(y)` is already predefined and will create the appropriate accessor function for you.

```
# Define custom accessor function
get_class <- function(x) attr(x, "class")
pluck(mtcars, get_class)
#> [1] "data.frame"

# Use attr_getter() as a helper
pluck(mtcars, attr_getter("class"))
#> [1] "data.frame"
```

Q2: `map(1:3, ~ runif(2))` is a useful pattern for generating random numbers, but `map(1:3, runif(2))` is not. Why not? Can you explain why it returns the result that it does?

A: The first pattern creates multiple random numbers, because `~ runif(2)` successfully uses the formula interface. Internally `map()` applies `as_mapper()` to this formula, which converts `~ runif(2)` into an anonymous function. Afterwards `runif(2)` is applied three times (one time during each iteration), leading to three different pairs of random numbers.

In the second pattern `runif(2)` is evaluated once, then the results are passed to `map()`. Consequently `as_mapper()` creates an extractor function based on the return values from `runif(2)` (via `pluck()`). This leads to three `NULL`s (`pluck()`'s `.default` return), because no values corresponding to the index can be found.

```
as_mapper(~ runif(2))
#> <lambda>
#> function (..., .x = ..1, .y = ..2, . = ..1)
#> runif(2)
#> attr(,"class")
#> [1] "rlang_lambda_function" "function"
as_mapper(runif(2))
#> function (x, ...)
#> pluck(x, 0.0807501375675201, 0.834333037259057, .default = NULL)
#> <environment: 0x55adc3b64390>
```

Q3: Use the appropriate `map()` function to:

a) Compute the standard deviation of every column in a numeric data frame.
b) Compute the standard deviation of every numeric column in a mixed data frame. (Hint: you'll need to do it in two steps.)
c) Compute the number of levels for every factor in a data frame.

A: To solve this exercise we take advantage of calling the type stable variants of `map()`, which give us more concise output, and use `map_lgl()` to select the columns of the data frame (later you'll learn about `keep()`, which simplifies this pattern a little).

```
map_dbl(mtcars, sd)
#>     mpg     cyl    disp      hp    drat      wt    qsec      vs
#>   6.027   1.786 123.939  68.563   0.535   0.978   1.787   0.504
#>      am    gear    carb
#>   0.499   0.738   1.615

penguins <- palmerpenguins::penguins

penguins_numeric <- map_lgl(penguins, is.numeric)
map_dbl(penguins[penguins_numeric], sd, na.rm = TRUE)
#>    bill_length_mm     bill_depth_mm flipper_length_mm
#>             5.460             1.975            14.062
#>       body_mass_g              year
#>           801.955             0.818

penguins_factor <- map_lgl(penguins, is.factor)
map_int(penguins[penguins_factor], ~ length(levels(.x)))
```

```
#> species  island     sex
#>       3       3       2
```

Q4: The following code simulates the performance of a t-test for non-normal data. Extract the p-value from each test, then visualise.

```
trials <- map(1:100, ~ t.test(rpois(10, 10), rpois(10, 7)))
```

A: There are many ways to visualise this data. However, since there are only 100 data points, we choose a dot plot to visualise the distribution. (Unfortunately, {ggplot2}s `geom_dotplot()` doesn't compute proper counts as it was created to visualise distribution densities instead of frequencies, so a histogram would be a suitable alternative).

```
library(ggplot2)

df_trials <- tibble::tibble(p_value = map_dbl(trials, "p.value"))

df_trials %>%
  ggplot(aes(x = p_value, fill = p_value < 0.05)) +
  geom_dotplot(binwidth = .01) +  # geom_histogram() as alternative
  theme(
    axis.text.y = element_blank(),
    axis.ticks.y = element_blank(),
    legend.position = "top"
  )
```

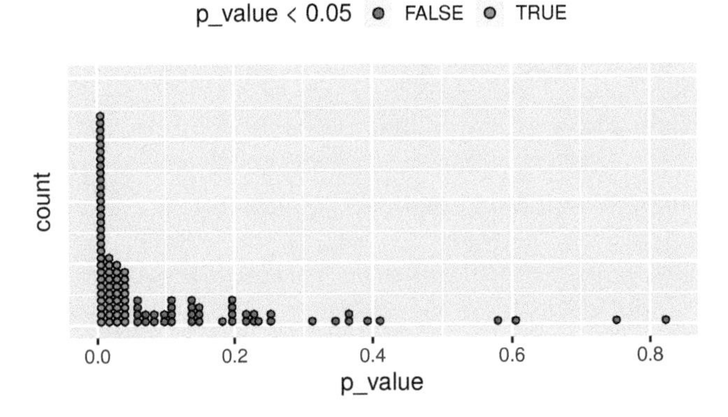

Q5: The following code uses a map nested inside another map to apply a function to every element of a nested list. Why does it fail, and what do you need to do to make it work?

```r
x <- list(
  list(1, c(3, 9)),
  list(c(3, 6), 7, c(4, 7, 6))
)

triple <- function(x) x * 3
map(x, map, .f = triple)
#> Error in .f(.x[[i]], ...): unused argument (function (.x, .f, ...)
#> {
#> .f <- as_mapper(.f, ...)
#> .Call(map_impl, environment(), ".x", ".f", "list")
#> })
```

A: This function call fails, because `triple()` is specified as the `.f` argument and consequently belongs to the outer `map()`. The unnamed argument `map` is treated as an argument of `triple()`, which causes the error.

There are a number of ways we could resolve the problem. However, there is not much to choose between them for this simple example, although it is good to know your options for more complicated cases.

```r
# Don't name the argument
map(x, map, triple)

# Use magrittr-style anonymous function
map(x, . %>% map(triple))

# Use purrr-style anonymous function
map(x, ~ map(.x, triple))
```

Q6: Use `map()` to fit linear models to the `mtcars` dataset using the formulas stored in this list:

```r
formulas <- list(
  mpg ~ disp,
  mpg ~ I(1 / disp),
  mpg ~ disp + wt,
  mpg ~ I(1 / disp) + wt
)
```

A: The data (`mtcars`) is constant for all these models and so we iterate over the `formulas` provided. As the formula is the first argument of `lm()`, we don't need to specify it explicitly.

```
models <- map(formulas, lm, data = mtcars)
```

Q7: Fit the model `mpg ~ disp` to each of the bootstrap replicates of `mtcars` in the list below, then extract the R^2 of the model fit (Hint: you can compute the R^2 with `summary()`)

```
bootstrap <- function(df) {
  df[sample(nrow(df), replace = TRUE), , drop = FALSE]
}

bootstraps <- map(1:10, ~ bootstrap(mtcars))
```

A: To accomplish this task, we take advantage of the "list in, list out"-functionality of `map()`. This allows us to chain multiple transformations together. We start by fitting the models. We then calculate the summaries and extract the R^2 values. For the last call we use `map_dbl()`, which provides convenient output.

```
bootstraps %>%
  map(~ lm(mpg ~ disp, data = .x)) %>%
  map(summary) %>%
  map_dbl("r.squared")
#>  [1] 0.588 0.822 0.745 0.746 0.784 0.749 0.613 0.792 0.653 0.726
```

9.4 Map variants

Q1: Explain the results of `modify(mtcars, 1)`.

A: `modify()` is based on `map()`, and in this case, the extractor interface will be used. It extracts the first element of each column in `mtcars`. `modify()` always returns the same structure as its input: in this case it forces the first row to be recycled 32 times. (Internally `modify()` uses `.x[] <- map(.x, .f, ...)` for assignment.)

```
head(modify(mtcars, 1))
#>                 mpg cyl disp  hp drat   wt qsec vs am gear carb
#> Mazda RX4        21   6  160 110  3.9 2.62 16.5  0  1    4    4
#> Mazda RX4 Wag    21   6  160 110  3.9 2.62 16.5  0  1    4    4
#> Datsun 710       21   6  160 110  3.9 2.62 16.5  0  1    4    4
```

```
#> Hornet 4 Drive      21   6   160 110   3.9 2.62 16.5  0  1    4    4
#> Hornet Sportabout  21   6   160 110   3.9 2.62 16.5  0  1    4    4
#> Valiant             21   6   160 110   3.9 2.62 16.5  0  1    4    4
```

Q2: Rewrite the following code to use `iwalk()` instead of `walk2()`. What are the advantages and disadvantages?

```
cyls <- split(mtcars, mtcars$cyl)
paths <- file.path(temp, paste0("cyl-", names(cyls), ".csv"))
walk2(cyls, paths, write.csv)
```

A: `iwalk()` allows us to use a single variable, storing the output path in the names.

```
temp <- tempfile()
dir.create(temp)

cyls <- split(mtcars, mtcars$cyl)
names(cyls) <- file.path(temp, paste0("cyl-", names(cyls), ".csv"))
iwalk(cyls, ~ write.csv(.x, .y))
```

We could do this in a single pipe by taking advantage of `set_names()`:

```
mtcars %>%
  split(mtcars$cyl) %>%
  set_names(~ file.path(temp, paste0("cyl-", .x, ".csv"))) %>%
  iwalk(~ write.csv(.x, .y))
```

Q3: Explain how the following code transforms a data frame using functions stored in a list.

```
trans <- list(
  disp = function(x) x * 0.0163871,
  am = function(x) factor(x, labels = c("auto", "manual"))
)

nm <- names(trans)
mtcars[nm] <- map2(trans, mtcars[nm], function(f, var) f(var))
```

Compare and contrast the `map2()` approach to this `map()` approach:

```
mtcars[nm] <- map(nm, ~ trans[[.x]](mtcars[[.x]]))
```

A: In the first approach

```
mtcars[nm] <- map2(trans, mtcars[nm], function(f, var) f(var))
```

the list of the 2 functions (`trans`) and the 2 appropriately selected data frame columns (`mtcars[nm]`) are supplied to `map2()`. `map2()` creates an anonymous function (`f(var)`) which applies the functions to the variables when `map2()` iterates over their (similar) indices. On the left-hand side, the respective 2 elements of `mtcars` are being replaced by their new transformations.

The `map()` variant

```
mtcars[nm] <- map(nm, ~ trans[[.x]](mtcars[[.x]]))
```

does basically the same. However, it directly iterates over the names (`nm`) of the transformations. Therefore, the data frame columns are selected during the iteration.

Besides the iteration pattern, the approaches differ in the possibilities for appropriate argument naming in the `.f` argument. In the `map2()` approach we iterate over the elements of `x` and `y`. Therefore, it is possible to choose appropriate placeholders like `f` and `var`. This makes the anonymous function more expressive at the cost of making it longer. We think using the formula interface in this way is preferable compared to the rather cryptic `mtcars[nm] <- map2(trans, mtcars[nm], ~ .x(.y))`.

In the `map()` approach we map over the variable names. It is therefore not possible to introduce placeholders for the function and variable names. The formula syntax together with the `.x` pronoun is pretty compact. The object names and the brackets clearly indicate the application of transformations to specific columns of `mtcars`. In this case the iteration over the variable names comes in handy, as it highlights the importance of matching between `trans` and `mtcars` element names. Together with the replacement form on the left-hand side, this line is relatively easy to inspect.

To summarise, in situations where `map()` and `map2()` provide solutions for an iteration problem, several points may be considered before deciding for one or the other approach.

Q4: What does `write.csv()` return, i.e. what happens if you use it with `map2()` instead of `walk2()`?

A: `write.csv()` returns `NULL`. As we call the function for its side effect (creating a CSV file), `walk2()` would be appropriate here. Otherwise, we receive a rather uninformative list of `NULL`s.

```
cyls <- split(mtcars, mtcars$cyl)
paths <- file.path(tempdir(), paste0("cyl-", names(cyls), ".csv"))

map2(cyls, paths, write.csv)
#> $`4`
#> NULL
#>
#> $`6`
#> NULL
#>
#> $`8`
#> NULL
```

9.6 Predicate functionals

Q1: Why isn't `is.na()` a predicate function? What base R function is closest to being a predicate version of `is.na()`?

A: `is.na()` is not a predicate function, because it returns a logical *vector* the same length as the input, not a single `TRUE` or `FALSE`.

`anyNA()` is the closest equivalent because it always returns a single `TRUE` or `FALSE` if there are any missing values present. You could also imagine an `allNA()` which would return `TRUE` if all values were missing, but that's considerably less useful so base R does not provide it.

Q2: `simple_reduce()` has a problem when `x` is length 0 or length 1. Describe the source of the problem and how you might go about fixing it.

```
simple_reduce <- function(x, f) {
  out <- x[[1]]
  for (i in seq(2, length(x))) {
    out <- f(out, x[[i]])
  }
  out
}
```

A: The loop inside `simple_reduce()` always starts with the index 2, and `seq()` can count both up *and* down:

```
seq(2, 0)
#> [1] 2 1 0
seq(2, 1)
#> [1] 2 1
```

Therefore, subsetting length-0 and length-1 vectors via `[[` will lead to a *subscript out of bounds* error. To avoid this, we allow `simple_reduce()` to return before the for loop is started and include a default argument for 0-length vectors.

```
simple_reduce <- function(x, f, default) {
  if (length(x) == 0L) return(default)
  if (length(x) == 1L) return(x[[1L]])

  out <- x[[1]]
  for (i in seq(2, length(x))) {
    out <- f(out, x[[i]])
  }
  out
}
```

Our new `simple_reduce()` now works as intended:

```
simple_reduce(integer(0), `+`)
#> Error in simple_reduce(integer(0), `+`): argument "default" is
#> missing, with no default
simple_reduce(integer(0), `+`, default = 0L)
#> [1] 0
simple_reduce(1, `+`)
#> [1] 1
simple_reduce(1:3, `+`)
#> [1] 6
```

Q3: Implement the `span()` function from Haskell: given a list `x` and a predicate function `f`, `span(x, f)` returns the location of the longest sequential run of elements where the predicate is true. (Hint: you might find `rle()` helpful.)

A: Our `span_r()` function returns the indices of the (first occurring) longest sequential run of elements where the predicate is true. If the predicate is never true, the longest run has length 0, in which case we return `integer(0)`.

```
span_r <- function(x, f) {
  idx <- unname(map_lgl(x, ~ f(.x)))
  rle <- rle(idx)

  # Check if the predicate is never true
  if (!any(rle$values)) {
    return(integer(0))
  }

  # Find the length of the longest sequence of true values
  longest <- max(rle$lengths[rle$values])
  # Find the positition of the (first) longest run in rle
  longest_idx <- which(rle$values & rle$lengths == longest)[1]

  # Add up all lengths in rle before the longest run
  ind_before_longest <- sum(rle$lengths[seq_len(longest_idx - 1)])

  out_start <- ind_before_longest + 1L
  out_end <- ind_before_longest + longest
  out_start:out_end
}

# Check that it works
span_r(c(0,  0,  0,  0,  0), is.na)
#> integer(0)
span_r(c(NA, 0,  0,  0,  0), is.na)
#> [1] 1
span_r(c(NA, 0, NA, NA, NA), is.na)
#> [1] 3 4 5
```

Q4: Implement `arg_max()`. It should take a function and a vector of inputs, and return the elements of the input where the function returns the highest value. For example, `arg_max(-10:5, function(x) x ^ 2)` should return `-10`. `arg_max(-5:5, function(x) x ^ 2)` should return `c(-5, 5)`. Also implement the matching `arg_min()` function.

A: Both functions take a vector of inputs and a function as an argument. The function output is then used to subset the input accordingly.

```
arg_max <- function(x, f) {
  y <- map_dbl(x, f)
  x[y == max(y)]
}
```

```
arg_min <- function(x, f) {
  y <- map_dbl(x, f)
  x[y == min(y)]
}

arg_max(-10:5, function(x) x ^ 2)
#> [1] -10
arg_min(-10:5, function(x) x ^ 2)
#> [1] 0
```

Q5: The function below scales a vector so it falls in the range [0, 1]. How would you apply it to every column of a data frame? How would you apply it to every numeric column in a data frame?

```
scale01 <- function(x) {
  rng <- range(x, na.rm = TRUE)
  (x - rng[1]) / (rng[2] - rng[1])
}
```

A: To apply a function to every column of a data frame, we can use `purrr::modify()` (or `purrr::map_dfr()`), which also conveniently returns a data frame. To limit the application to numeric columns, the scoped version `modify_if()` can be used.

```
modify_if(mtcars, is.numeric, scale01)
```

9.7 Base functionals

Q1: How does `apply()` arrange the output? Read the documentation and perform some experiments.

A: Basically `apply()` applies a function over the margins of an array. In the two-dimensional case, the margins are just the rows and columns of a matrix. Let's make this concrete.

```
arr2 <- array(1:12, dim = c(3, 4))
rownames(arr2) <- paste0("row", 1:3)
colnames(arr2) <- paste0("col", 1:4)
arr2
```

```
#>      col1 col2 col3 col4
#> row1    1    4    7   10
#> row2    2    5    8   11
#> row3    3    6    9   12
```

When we apply the `head()` function over the first margin of `arr2()` (i.e. the rows), the results are contained in the columns of the output, transposing the array compared to the original input.

```
apply(arr2, 1, function(x) x[1:2])
#>      row1 row2 row3
#> col1    1    2    3
#> col2    4    5    6
```

And vice versa if we apply over the second margin (the columns):

```
apply(arr2, 2, function(x) x[1:2])
#>      col1 col2 col3 col4
#> row1    1    4    7   10
#> row2    2    5    8   11
```

The output of `apply()` is organised first by the margins being operated over, then the results of the function. This can become quite confusing for higher dimensional arrays.

Q2: What do `eapply()` and `rapply()` do? Does `{purrr}` have equivalents?

A: `eapply()` is a variant of `lapply()`, which iterates over the (named) elements of an environment. In `{purrr}` there is no equivalent for `eapply()` as `{purrr}` mainly provides functions that operate on vectors and functions, but not on environments.

`rapply()` applies a function to all elements of a list recursively. This function makes it possible to limit the application of the function to specified classes (default `classes = ANY`). One may also specify how elements of other classes should remain: as their identity (`how = replace`) or another value (`default = NULL`). The closest equivalent in `{purrr}` is `modify_depth()`, which allows you to modify elements at a specified depth in a nested list.

Q3: Challenge: read about the fixed point algorithm (`https://mitpress.mit.edu/sites/default/files/sicp/full-text/book/book-Z-H-12.html#%25_idx_1096`). Complete the exercises using R.

A: A number x is called a fixed point of a function f if it satisfies the equation $f(x) = x$. For some functions we may find a fixed point by beginning with a starting value and applying f repeatedly. Here `fixed_point()` acts as a functional because it takes a function as an argument.

```
fixed_point <- function(f, x_init, n_max = 10000, tol = 0.0001) {
  n <- 0
  x <- x_init
  y <- f(x)

  is_fixed_point <- function(x, y) {
    abs(x - y) < tol
  }

  while (!is_fixed_point(x, y)) {
    x <- y
    y <- f(y)

    # Make sure we eventually stop
    n <- n + 1
    if (n > n_max) {
      stop("Failed to converge.", call. = FALSE)
    }
  }

  x
}


# Functions with fixed points
fixed_point(sin, x_init = 1)
#> [1] 0.0843
fixed_point(cos, x_init = 1)
#> [1] 0.739

# Functions without fixed points
add_one <- function(x) x + 1
fixed_point(add_one, x_init = 1)
#> Error: Failed to converge.
```

10

Function factories

Prerequisites

For most of this chapter base R [R Core Team, 2020] is sufficient. Just a few exercises require the `{rlang}` [Henry and Wickham, 2020b], `{dplyr}` [Wickham et al., 2020a], `{purrr}` [Henry and Wickham, 2020a] and `{ggplot2}` [Wickham, 2016] packages.

```
library(rlang)
library(dplyr)
library(purrr)
library(ggplot2)
```

10.2 Factory fundamentals

Q1: The definition of `force()` is simple:

```
force
#> function (x)
#> x
#> <bytecode: 0x5609b8696740>
#> <environment: namespace:base>
```

Why is it better to `force(x)` instead of just `x`?

A: As you can see `force(x)` is similar to `x`. As mentioned in *Advanced R*, we prefer this explicit form, because

> using this function clearly indicates that you're forcing evaluation, not that you've accidentally typed `x`."

DOI: 10.1201/9781003175414-10

Q2: Base R contains two function factories, `approxfun()` and `ecdf()`. Read their documentation and experiment to figure out what the functions do and what they return.

A: Let's begin with `approxfun()` as it is used within `ecdf()` as well:

`approxfun()` takes a combination of data points (x and y values) as input and returns a stepwise linear (or constant) interpolation function. To find out what this means exactly, we first create a few random data points.

```
x <- runif(10)
y <- runif(10)
plot(x, y, lwd = 10)
```

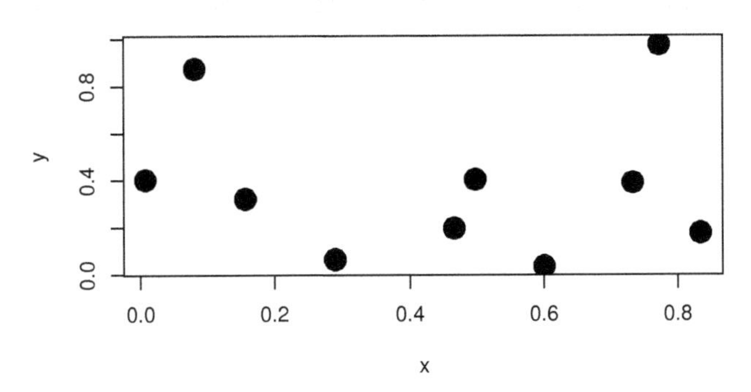

Next, we use `approxfun()` to construct the linear and constant interpolation functions for our `x` and `y` values.

```
f_lin <- approxfun(x, y)
f_con <- approxfun(x, y, method = "constant")

# Both functions exactly reproduce their input y values
identical(f_lin(x), y)
#> [1] TRUE
identical(f_con(x), y)
#> [1] TRUE
```

When we apply these functions to new x values, these are mapped to the lines connecting the initial y values (linear case) or to the same y value as for the next smallest initial x value (constant case).

```
x_new <- runif(1000)

plot(x, y, lwd = 10)
points(x_new, f_lin(x_new), col = "cornflowerblue", pch = 16)
points(x_new, f_con(x_new), col = "firebrick", pch = 16)
```

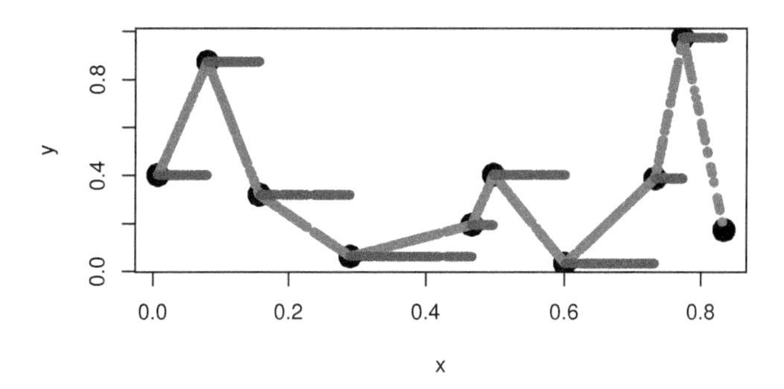

However, both functions are only defined within `range(x)`.

```
f_lin(range(x))
#> [1] 0.402 0.175
f_con(range(x))
#> [1] 0.402 0.175

(eps <- .Machine$double.neg.eps)
#> [1] 1.11e-16

f_lin(c(min(x) - eps, max(x) + eps))
#> [1] NA NA
f_con(c(min(x) - eps, max(x) + eps))
#> [1] NA NA
```

To change this behaviour, one can set `rule = 2`. This leads to the result that for values outside of `range(x)` the boundary values of the function are returned.

```
f_lin <- approxfun(x, y, rule = 2)
f_con <- approxfun(x, y, method = "constant", rule = 2)

f_lin(c(-Inf, Inf))
#> [1] 0.402 0.175
f_con(c(-Inf, Inf))
#> [1] 0.402 0.175
```

Another option is to customise the return values as individual constants for each side via `yleft` and/or `yright`.

```
f_lin <- approxfun(x, y, yleft = 5)
f_con <- approxfun(x, y, method = "constant", yleft = 5, yright = -5)
```

```
f_lin(c(-Inf, Inf))
#> [1]  5 NA
f_con(c(-Inf, Inf))
#> [1]  5 -5
```

Further, approxfun() provides the option to shift the y values for method = "constant" between their left and right values. According to the documentation this indicates a compromise between left- and right-continuous steps.

```
f_con <- approxfun(x, y, method = "constant", f = .5)

plot(x, y, lwd = 10)
points(x_new, f_con(x_new), pch = 16)
```

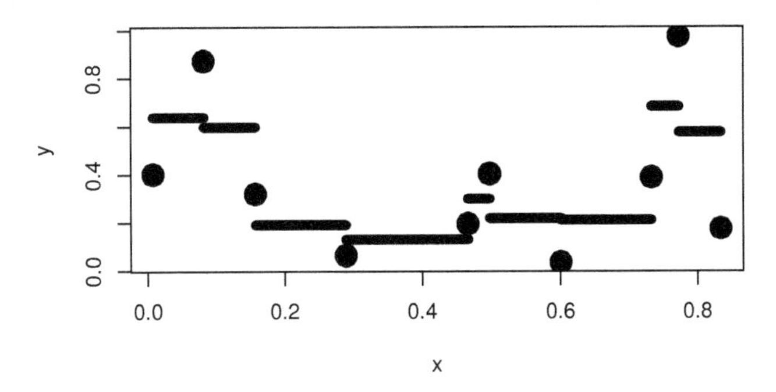

Finally, the ties argument allows to aggregate y values if multiple ones were provided for the same x value. For example, in the following line we use mean() to aggregate these y values before they are used for the interpolation approxfun(x = c(1,1,2), y = 1:3, ties = mean).

Next, we focus on ecdf(). "ecdf" is an acronym for empirical cumulative distribution function. For a numeric vector of density values, ecdf() initially creates the (x, y) pairs for the nodes of the density function and then passes these pairs to approxfun(), which gets called with specifically adapted settings (approxfun(vals, cumsum(tabulate(match(x, vals)))/n, method = "constant", yleft = 0, yright = 1, f = 0, ties = "ordered")).

```
x <- runif(10)
f_ecdf <- ecdf(x)
class(f_ecdf)
#> [1] "ecdf"     "stepfun"  "function"

plot(x, f_ecdf(x), lwd = 10, ylim = 0:1)
```

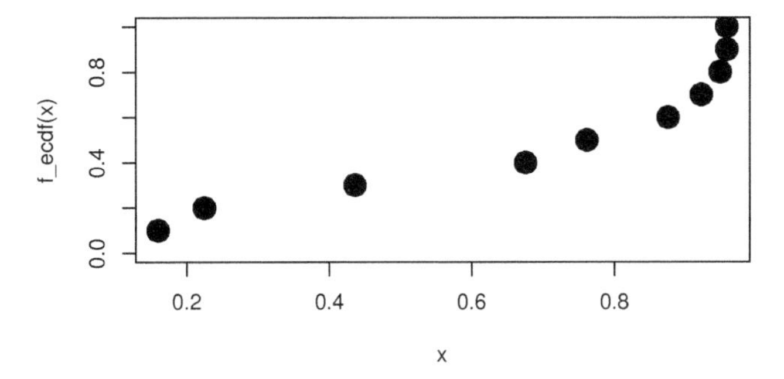

New values are then mapped on the y value of the next smallest x value from within the initial input.

```
x_new <- runif(1000)

plot(x, f_ecdf(x), lwd = 10, ylim = 0:1)
points(x_new, f_ecdf(x_new), ylim = 0:1)
```

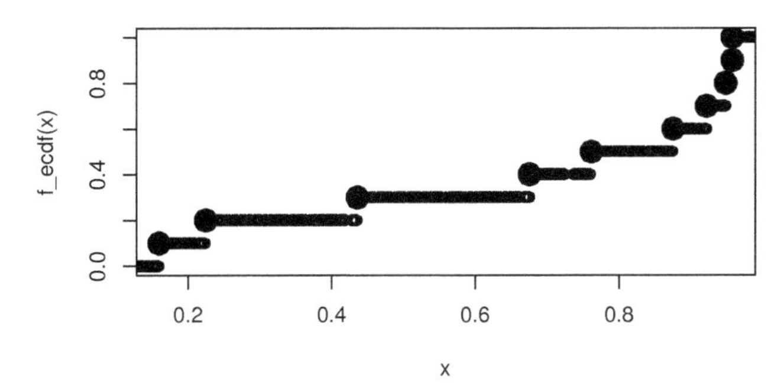

Q3: Create a function `pick()` that takes an index, `i`, as an argument and returns a function with an argument `x` that subsets `x` with `i`.

```
pick(1)(x)
# should be equivalent to
x[[1]]

lapply(mtcars, pick(5))
# should be equivalent to
lapply(mtcars, function(x) x[[5]])
```

A: In this exercise `pick(i)` acts as a function factory, which returns the required subsetting function.

```
pick <- function(i) {
  force(i)
```

```
  function(x) x[[i]]
}

x <- 1:3
identical(x[[1]], pick(1)(x))
#> [1] TRUE
identical(
  lapply(mtcars, function(x) x[[5]]),
  lapply(mtcars, pick(5))
)
#> [1] TRUE
```

Q4: Create a function that creates functions that compute the ith central moment (http://en.wikipedia.org/wiki/Central_moment) of a numeric vector. You can test it by running the following code:

```
m1 <- moment(1)
m2 <- moment(2)

x <- runif(100)
stopifnot(all.equal(m1(x), 0))
stopifnot(all.equal(m2(x), var(x) * 99 / 100))
```

A: The first moment is closely related to the mean and describes the average deviation from the mean, which is 0 (within numerical margin of error). The second moment describes the variance of the input data. If we want to compare it to `var()`, we need to undo Bessel's correction (https://en.wikipedia.org/wiki/Bessel%27s_correction) by multiplying with $\frac{N-1}{N}$.

```
moment <- function(i) {
  force(i)

  function(x) sum((x - mean(x)) ^ i) / length(x)
}

m1 <- moment(1)
m2 <- moment(2)

x <- runif(100)
all.equal(m1(x), 0)  # removed stopifnot() for clarity
#> [1] TRUE
```

```
all.equal(m2(x), var(x) * 99 / 100)
#> [1] TRUE
```

Q5: What happens if you don't use a closure? Make predictions, then verify with the code below.

```
i <- 0
new_counter2 <- function() {
  i <<- i + 1
  i
}
```

A: Without the captured and encapsulated environment of a closure the counts will be stored in the global environment. Here they can be overwritten or deleted as well as interfere with other counters.

```
new_counter2()
#> [1] 1
i
#> [1] 1
new_counter2()
#> [1] 2
i
#> [1] 2

i <- 0
new_counter2()
#> [1] 1
i
#> [1] 1
```

Q6: What happens if you use <- instead of <<-? Make predictions, then verify with the code below.

```
new_counter3 <- function() {
  i <- 0
  function() {
    i <- i + 1
    i
  }
}
```

A: Without the super assignment `<<-`, the counter will always return 1. The counter always starts in a new execution environment within the same enclosing environment, which contains an unchanged value for `i` (in this case it remains 0).

```
new_counter_3 <- new_counter3()

new_counter_3()
#> [1] 1
new_counter_3()
#> [1] 1
```

10.3 Graphical factories

Q1: Compare and contrast `ggplot2::label_bquote()` with `scales::number_format()`.

A: Both functions will help you in styling your output, e.g. in your plots and they do this by returning the desired formatting function to you.

`ggplot2::label_bquote()` takes relatively straightforward plotmath (`https://stat.ethz.ch/R-manual/R-patched/library/grDevices/html/plotmath.html`) expressions and uses them for faceting labels in `{ggplot2}`. Because this function is used in `{ggplot2}` it needs to return a function of `class = "labeller"`.

`scales::number_format()` initially `force()`s the computation of all parameters. It's essentially a parametrised wrapper around `scales::number()` and will help you format numbers appropriately. It will return a simple function.

10.4 Statistical factories

Q1: In `boot_model()`, why don't I need to force the evaluation of `df` or `model`?

A: `boot_model()` ultimately returns a function, and whenever you return a function you need to make sure all the inputs are explicitly evaluated. Here that happens automatically because we use `df` and `formula` in `lm()` before returning the function.

```
boot_model <- function(df, formula) {
  mod <- lm(formula, data = df)
  fitted <- unname(fitted(mod))
  resid <- unname(resid(mod))
  rm(mod)

  function() {
    fitted + sample(resid)
  }
}
```

Q2: Why might you formulate the Box-Cox transformation like this?

```
boxcox3 <- function(x) {
  function(lambda) {
    if (lambda == 0) {
      log(x)
    } else {
      (x ^ lambda - 1) / lambda
    }
  }
}
```

A: `boxcox3()` returns a function where `x` is fixed (though it is not forced, so it may be manipulated later). This allows us to apply and test different transformations for different inputs and give them a descriptive name.

```
boxcox_airpassengers <- boxcox3(AirPassengers)

plot(boxcox_airpassengers(0))
plot(boxcox_airpassengers(1))
plot(boxcox_airpassengers(2))
plot(boxcox_airpassengers(3))
```

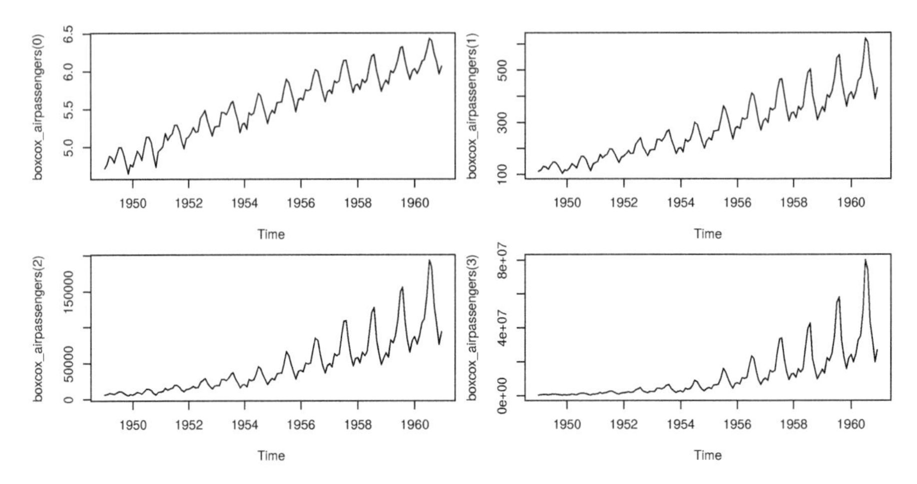

Q3: Why don't you need to worry that `boot_permute()` stores a copy of the data inside the function that it generates?

A: `boot_permute()` is defined in *Advanced R* as:

```
boot_permute <- function(df, var) {
  n <- nrow(df)
  force(var)

  function() {
    col <- df[[var]]
    col[sample(n, replace = TRUE)]
  }
}
```

We don't need to worry that it stores a copy of the data, because it actually doesn't store one; it's just a name that points to the same underlying object in memory.

```
boot_mtcars1 <- boot_permute(mtcars, "mpg")

lobstr::obj_size(mtcars)
#> 7,208 B
lobstr::obj_size(boot_mtcars1)
#> 20,232 B
lobstr::obj_sizes(mtcars, boot_mtcars1)
#> *  7,208 B
#> * 13,024 B
```

Q4: How much time does `ll_poisson2()` save compared to `ll_poisson1()`? Use `bench::mark()` to see how much faster the optimisation occurs. How does changing the length of `x` change the results?

A: Let us recall the definitions of `ll_poisson1()`, `ll_poisson2()` and the test data `x1`:

```
ll_poisson1 <- function(x) {
  n <- length(x)

  function(lambda) {
    log(lambda) * sum(x) - n * lambda - sum(lfactorial(x))
  }
}

ll_poisson2 <- function(x) {
  n <- length(x)
  sum_x <- sum(x)
  c <- sum(lfactorial(x))

  function(lambda) {
    log(lambda) * sum_x - n * lambda - c
  }
}

x1 <- c(41, 30, 31, 38, 29, 24, 30, 29, 31, 38)
```

A benchmark on `x1` reveals a performance improvement of factor 2 for `ll_poisson2()` over `ll_poisson1()`:

```
bench::mark(
  llp1 = optimise(ll_poisson1(x1), c(0, 100), maximum = TRUE),
  llp2 = optimise(ll_poisson2(x1), c(0, 100), maximum = TRUE)
)
#> # A tibble: 2 x 6
#>   expression      min   median `itr/sec` mem_alloc `gc/sec`
#>   <bch:expr> <bch:tm> <bch:tm>     <dbl> <bch:byt>    <dbl>
#> 1 llp1         31.8µs   36.7µs    26540.    12.8KB     29.2
#> 2 llp2           16µs   18.2µs    52523.        0B     26.3
```

As the redundant calculations within `ll_poisson1()` become more expensive with growing length of `x1`, we expect even further relative performance improvements for `ll_poisson2()`. The following benchmark reveals a relative performance improvement of factor 20 for `ll_poisson2()` when `x1` is of length 100,000:

```
bench_poisson <- function(x_length) {
  x <- rpois(x_length, 100L)

  bench::mark(
    llp1 = optimise(ll_poisson1(x), c(0, 100), maximum = TRUE),
    llp2 = optimise(ll_poisson2(x), c(0, 100), maximum = TRUE),
    time_unit = "ms"
  )
}

performances <- map_dfr(10^(1:5), bench_poisson)

df_perf <- tibble(
  x_length = rep(10^(1:5), each = 2),
  method   = rep(attr(performances$expression, "description"), 5),
  median   = performances$median
)

ggplot(df_perf, aes(x_length, median, col = method)) +
  geom_point(size = 2) +
  geom_line(linetype = 2) +
  scale_x_log10() +
  labs(
    x = "Length of x",
    y = "Execution Time (ms)",
    color = "Method"
  ) +
  theme(legend.position = "top")
```

Method
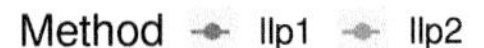

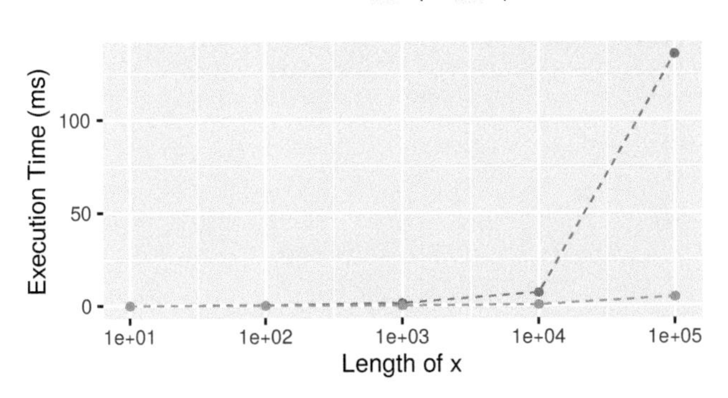

10.5 Function factories + functionals

Q1: Which of the following commands is equivalent to `with(x, f(z))`?

(a) `x$f(x$z)`.
(b) `f(x$z)`.
(c) `x$f(z)`.
(d) `f(z)`.
(e) It depends.

A: (e) "It depends" is the correct answer. Usually `with()` is used with a data frame, so you'd usually expect (b), but if `x` is a list, it could be any of the options.

```
f <- mean
z <- 1
x <- list(f = mean, z = 1)

identical(with(x, f(z)), x$f(x$z))
#> [1] TRUE
identical(with(x, f(z)), f(x$z))
#> [1] TRUE
identical(with(x, f(z)), x$f(z))
#> [1] TRUE
identical(with(x, f(z)), f(z))
#> [1] TRUE
```

Q2: Compare and contrast the effects of `env_bind()` vs. `attach()` for the following code.

```
funs <- list(
  mean = function(x) mean(x, na.rm = TRUE),
  sum = function(x) sum(x, na.rm = TRUE)
)

attach(funs)
#> The following objects are masked from package:base:
#>
#>     mean, sum
mean <- function(x) stop("Hi!")
detach(funs)
```

```
env_bind(globalenv(), !!!funs)
mean <- function(x) stop("Hi!")
env_unbind(globalenv(), names(funs))
```

A: `attach()` adds `funs` to the search path. Therefore, the provided functions are found before their respective versions from the `{base}` package. Further, they cannot get accidentally overwritten by similar named functions in the global environment. One annoying downside of using `attach()` is the possibility to attach the same object multiple times, making it necessary to call `detach()` equally often.

```
attach(funs)
#> The following objects are masked from package:base:
#>
#>     mean, sum
attach(funs)
#> The following objects are masked from funs (pos = 3):
#>
#>     mean, sum
#>
#> The following objects are masked from package:base:
#>
#>     mean, sum
```

```
head(search())
#> [1] ".GlobalEnv"       "funs"             "funs"
#> [4] "package:ggplot2" "package:purrr"    "package:dplyr"
detach(funs)
detach(funs)
```

In contrast `rlang::env_bind()` just adds the functions in `fun` to the global environment. No further side effects are introduced, and the functions are overwritten when similarly named functions are defined.

```
env_bind(globalenv(), !!!funs)
head(search())
#> [1] ".GlobalEnv"       "package:ggplot2" "package:purrr"
#> [4] "package:dplyr"    "package:rlang"    "package:stats"
```

11

Function operators

Prerequisites

Also in the third chapter on functional programming, we make relatively frequent use of the {purrr} package.

```
library(purrr)
```

11.2 Existing function operators

Q1: Base R provides a function operator in the form of `Vectorize()`. What does it do? When might you use it?

A: In R a lot of functions are "vectorised". Vectorised has two meanings. First, it means (broadly) that a function inputs a vector or vectors and does something to each element. Secondly, it usually implies that these operations are implemented in a compiled language such as C or Fortran, so that the implementation is very fast.

However, despite what the function's name implies, `Vectorize()` is not able to speed up the provided function. It rather changes the input format of the supplied arguments (`vectorize.args`), so that they can be iterated over.

Let's take a look at an example from the documentation:

```
vrep <- Vectorize(rep.int)
vrep
#> function (x, times)
#> {
#>   args <- lapply(as.list(match.call())[-1L], eval, parent.frame())
```

DOI: 10.1201/9781003175414-11

```
#>   names <- if (is.null(names(args)))
#>     character(length(args))
#>   else names(args)
#>   dovec <- names %in% vectorize.args
#>   do.call("mapply", c(FUN = FUN, args[dovec],
#>                        MoreArgs = list(args[!dovec]),
#>                        SIMPLIFY = SIMPLIFY, USE.NAMES = USE.NAMES))
#> }
#> <environment: 0x558902db65d0>
```

```
# Application
vrep(1:2, 3:4)
#> [[1]]
#> [1] 1 1 1
#>
#> [[2]]
#> [1] 2 2 2 2
```

`Vectorize()` provides a convenient and concise notation to iterate over multiple arguments but has some major drawbacks that mean you generally shouldn't use it. See `https://www.jimhester.com/post/2018-04-12-vectorize/` for more details.

Q2: Read the source code for `possibly()`. How does it work?

A: `possibly()` modifies functions to return a specified default value (`otherwise`) in case of an error and to suppress any error messages (`quiet = TRUE`).

While reading the source code, we notice that `possibly()` internally uses `purrr::as_mapper()`. This enables users to supply not only functions, but also formulas or atomics via the same syntax as known from other functions in the `{purrr}` package. Besides this, the new default value (`otherwise`) gets evaluated once to make it (almost) immutable.

```
possibly
#> function (.f, otherwise, quiet = TRUE)
#> {
#>     .f <- as_mapper(.f)
#>     force(otherwise)
#>     function(...) {
#>         tryCatch(.f(...), error = function(e) {
#>             if (!quiet)
#>                 message("Error: ", e$message)
#>             otherwise
```

```
#>         }, interrupt = function(e) {
#>             stop("Terminated by user", call. = FALSE)
#>         })
#>     }
#> }
#> <bytecode: 0x563747a5ff68>
#> <environment: namespace:purrr>
```

The main functionality of `possibly()` is provided by `base::tryCatch()`. In this part the supplied function (`.f`) gets wrapped and the error and interrupt handling are specified.

Q3: Read the source code for `safely()`. How does it work?

A: `safely()` modifies functions to return a list, containing the elements `result` and `error`. It works in a similar fashion as `possibly()` and besides using `as_mapper()`, `safely()` also provides the `otherwise` and `quiet` arguments. However, in order to provide the result and the error in a consistent way, the `tryCatch()` part of the implementation returns a list with similar structure for both cases. In the case of successful evaluation `error` equals `NULL` and in case of an error `result` equals `otherwise`, which is `NULL` by default.

As the `tryCatch()` part is hidden in the internal `purrr:::capture_output()` function, we provide it here in addition to `safely()`:

```
safely
#> function (.f, otherwise = NULL, quiet = TRUE)
#> {
#>     .f <- as_mapper(.f)
#>     function(...) capture_error(.f(...), otherwise, quiet)
#> }
#> <bytecode: 0x56374811ce48>
#> <environment: namespace:purrr>
```

```
purrr:::capture_error
#> function (code, otherwise = NULL, quiet = TRUE)
#> {
#>     tryCatch(list(result = code, error = NULL), error = function(e) {
#>         if (!quiet)
#>             message("Error: ", e$message)
#>         list(result = otherwise, error = e)
#>     }, interrupt = function(e) {
#>         stop("Terminated by user", call. = FALSE)
#>     })
#> }
```

```
#> <bytecode: 0x5637481aeb48>
#> <environment: namespace:purrr>
```

Take a look at *Advanced R* or the documentation of `safely()` to see how you can take advantage of this behaviour, e.g. when fitting many models.

11.3 Case study: Creating your own function operators

Q1: Weigh the pros and cons of `download.file %>% dot_every(10) %>% delay_by(0.1)` versus `download.file %>% delay_by(0.1) %>% dot_every(10)`.

A: Both commands will print a dot every 10 downloads and will take the same amount of time to run, so the differences may seem quite subtle.

In the first case, first the dot functionality is added to `download.file()`. Then the delay is added to this already tweaked function. This implies, that the printing of the dot will also be delayed, and the first dot will be printed as soon as the download for the 10th URL starts.

In the latter case the delay is added first and the dot-functionality is wrapped around it. This order will print the first dot immediately after the 9th download is finished, then the short delay occurs before the 10th download actually starts.

Q2: Should you memoise `file.download()`? Why or why not?

A: Memoising `file.download()` will only work if the files are immutable, i.e. if the file at a given URL is always the same. There's no point memoising unless this is true. Even if this is true, however, memoise has to store the results in memory, and large files will potentially take up a lot of memory.

This implies that it's probably not beneficial to memoise `file.download()` in most cases. The only exception is if you are downloading small files many times, and the file at a given URL is guaranteed not to change.

Q3: Create a function operator that reports whenever a file is created or deleted in the working directory, using `dir()` and `setdiff()`. What other global function effects might you want to track?

A: We start with a function that reports the difference between two vectors containing file names:

```
dir_compare <- function(old, new) {
  if (setequal(old, new)) {
    return()
  }

  added <- setdiff(new, old)
  removed <- setdiff(old, new)

  changes <- c(
    if (length(added) > 0) paste0(" * '", added, "' was added"),
    if (length(removed) > 0) paste0(" * '", removed ,
                                    "' was removed")
  )
  message(paste(changes, collapse = "\n"))
}

dir_compare(c("x", "y"), c("x", "y"))
#> NULL
dir_compare(c("x", "y"), c("x", "a"))
#>  * 'a' was added
#>  * 'y' was removed
```

Then we wrap it up in a function operator

```
track_dir <- function(f) {
  force(f)
  function(...) {
    dir_old <- dir()
    on.exit(dir_compare(dir_old, dir()), add = TRUE)

    f(...)
  }
}
```

And try it out by creating wrappers around `file.create()` and `file.remove()`:

```
file_create <- track_dir(file.create)
file_remove <- track_dir(file.remove)

file_create("delete_me")
#>  * 'delete_me' was added
#> [1] TRUE
file_remove("delete_me")
```

```
#>  * 'delete_me' was removed
#> [1] TRUE
```

To create a more serious version of `track_dir()` one might provide optionality to set the `full.names` and `recursive` arguments of `dir()` to `TRUE`. This would enable to also track the creation/deletion of hidden files and files in folders contained in the working directory.

Other global effects that might be worth tracking include changes regarding:

- the search path and possibly introduced `conflicts()`
- `options()` and `par()` which modify global settings
- the path of the working directory
- environment variables

Q4: Write a function operator that logs a timestamp and message to a file every time a function is run.

A: Our `logger()` function operator takes a function and a file path as input. One timestamp is written to the file under `log_path` when we call `logger()` and another timestamp is written to the same file each time the new function gets called.

```
append_line <- function(path, ...) {
  cat(..., "\n", sep = "", file = path, append = TRUE)
}

logger <- function(f, log_path) {
  force(f)
  force(log_path)

  append_line(log_path, "created at: ", as.character(Sys.time()))
  function(...) {
    append_line(log_path, "called at: ", as.character(Sys.time()))
    f(...)
  }
}
```

Now, let's check if our `logger()` works as intended and apply it to the `mean()` function:

```
log_path <- tempfile()
mean2 <- logger(mean, log_path)
Sys.sleep(5)
mean2(1:4)
```

```
#> [1] 2.5
Sys.sleep(1)
mean2(1:4)
#> [1] 2.5

readLines(log_path)
#> [1] "created at: 2020-12-27 23:30:26"
#> [2] "called at: 2020-12-27 23:30:31"
#> [3] "called at: 2020-12-27 23:30:32"
```

Q5: Modify `delay_by()` so that instead of delaying by a fixed amount of time, it ensures that a certain amount of time has elapsed since the function was last called. That is, if you called `g <- delay_by(1, f); g(); Sys.sleep(2); g()` there shouldn't be an extra delay.

A: `delay_by()` was defined in *Advanced R* as:

```
delay_by <- function(f, amount) {
  force(f)
  force(amount)

  function(...) {
    Sys.sleep(amount)
    f(...)
  }
}
```

To ensure that the function created by `delay_by()` waits that a certain amount of time has passed since its last execution, we incorporate three little changes into our new `delay_atleast()` as indicated in the corresponding comments below.

```
delay_atleast <- function(amount, f) {
  force(f)
  force(amount)

  # Store the last time the function was run
  last_time <- NULL

  # Return modified "delay-aware" function
  function(...) {
    if (!is.null(last_time)) {
      wait <- (last_time - Sys.time()) + amount
```

```
      if (wait > 0) {
        Sys.sleep(wait)
      }
    }

    # Update the time after the function has finished
    on.exit(last_time <<- Sys.time())

    f(...)
  }
}
```

Part III

Object-oriented programming

13

S3

Prerequisites

To interact with S3 objects, we will mainly use the `{sloop}` package [Wickham, 2019b].

```
library(sloop)
```

13.2 Basics

Q1: Describe the difference between `t.test()` and `t.data.frame()`? When is each function called?

A: Because of S3's `generic.class()` naming scheme, both functions may initially look similar, while they are in fact unrelated.

- `t.test()` is a *generic* function that performs a t-test.
- `t.data.frame()` is a *method* that gets called by the generic `t()` to transpose data frame input.

Due to R's S3 dispatch rules, `t.test()` would also get called when `t()` is applied to an object of class `test`.

Q2: Make a list of commonly used base R functions that contain `.` in their name but are not S3 methods.

A: In recent years "snake_case"-style has become increasingly common when naming functions and variables in R. But many functions in base R will continue to be "point.separated", which is why some inconsistency in your R code most likely cannot be avoided.[Bååth, 2012]

DOI: 10.1201/9781003175414-13

```
# Some base R functions with point.separated names
install.packages()
read.csv()

list.files()
download.file()

data.frame()
as.character()
Sys.Date()

all.equal()

do.call()
on.exit()
```

Q3: What does the `as.data.frame.data.frame()` method do? Why is it confusing? How could you avoid this confusion in your own code?

A: The function `as.data.frame.data.frame()` implements the `data.frame()` *method* for the `as.data.frame()` *generic*, which coerces objects to data frames.

The name is confusing, because it does not clearly communicate the type of the function, which could be a regular function, a generic or a method. Even if we assume a method, the amount of .'s makes it difficult to separate the generic- and the class-part of the name. Is it the `data.frame.data.frame()` method for the `as()` generic? Is it the `frame.data.frame()` method for the `as.data()` generic?

We could avoid this confusion by applying a different naming convention (e.g. "snake_case") for our class and function names.

Q4: Describe the difference in behaviour in these two calls.

```
some_days <- as.Date("2017-01-31") + sample(10, 5)

mean(some_days)
#> [1] "2017-02-06"
mean(unclass(some_days))
#> [1] 17203
```

A: `mean()` is a generic function, which will select the appropriate method based on the class of the input. `some_days` has the class `Date` and `mean.Date(some_days)` will be used to calculate the mean date of `some_days`.

After `unclass()` has removed the class attribute from `some_date`, the default method is chosen. `mean.default(unclass(some_days))` then calculates the mean of the underlying double.

Q5: What class of object does the following code return? What base type is it built on? What attributes does it use?

```
x <- ecdf(rpois(100, 10))
x
#> Empirical CDF
#> Call: ecdf(rpois(100, 10))
#>  x[1:18] =  2,  3,  4,  ..., 2e+01, 2e+01
```

A: It returns an object of the class `ecdf` (empirical cumulative distribution function) with the superclasses `stepfun` and `function`. The `ecdf` object is built on the base type `closure` (a function). The expression, which was used to create it (`rpois(100, 10)`), is stored in the `call` attribute.

```
typeof(x)
#> [1] "closure"

attributes(x)
#> $class
#> [1] "ecdf"     "stepfun"  "function"
#> 
#> $call
#> ecdf(rpois(100, 10))
```

Q6: What class of object does the following code return? What base type is it built on? What attributes does it use?

```
x <- table(rpois(100, 5))
x
#> 
#>  1  2  3  4  5  6  7  8  9 10 
#>  7  5 18 14 15 15 14  4  5  3
```

A: This code returns a `table` object, which is built upon the `integer` type. The attribute `dimnames` is used to name the elements of the integer vector.

```
typeof(x)
#> [1] "integer"
```

```
attributes(x)
#> $dim
#> [1] 10
#>
#> $dimnames
#> $dimnames[[1]]
#>  [1] "1"  "2"  "3"  "4"  "5"  "6"  "7"  "8"  "9"  "10"
#>
#>
#> $class
#> [1] "table"
```

13.3 Classes

Q1: Write a constructor for `data.frame` objects. What base type is a data frame built on? What attributes does it use? What are the restrictions placed on the individual elements? What about the names?

A: Data frames are built on named lists of vectors, which all have the same length. Besides the `class` and the column names (`names`), the `row.names` are their only further attribute. This must be a character vector with the same length as the other vectors.

We need to provide the number of rows as an input to make it possible to create data frames with 0 columns but multiple rows.

This leads to the following constructor:

```
new_data.frame <- function(x, n, row.names = NULL) {
  # Check if the underlying object is a list
  stopifnot(is.list(x))

  # Check all inputs are the same length
  # (This check also allows that x has length 0)
  stopifnot(all(lengths(x) == n))

  if (is.null(row.names)) {
    # Use special row names helper from base R
    row.names <- .set_row_names(n)
  } else {
```

```
    # Otherwise check that they're a character vector with the
    # correct length
    stopifnot(is.character(row.names), length(row.names) == n)
  }

  structure(
    x,
    class = "data.frame",
    row.names = row.names
  )
}

# Test
x <- list(a = 1, b = 2)
new_data.frame(x, n = 1)
#>   a b
#> 1 1 2
new_data.frame(x, n = 1, row.names = "l1")
#>    a b
#> l1 1 2

# Create a data frame with 0 columns and 2 rows
new_data.frame(list(), n = 2)
#> data frame with 0 columns and 2 rows
```

There are two additional restrictions we could implement if we were being very strict: both the row names and column names should be unique.

Q2: Enhance my `factor()` helper to have better behaviour when one or more values is not found in `levels`. What does `base::factor()` do in this situation?

A: `base::factor()` converts these values (silently) into `NA`s:

```
factor(c("a", "b", "c"), levels = c("a", "b"))
#> [1] a    b    <NA>
#> Levels: a b
```

The `factor()` helper including the constructor (`new_factor()`) and its validator (`validate_factor()`) were given in *Advanced R*. However, as the goal of this question is to throw an early error within the helper, we only repeat the code for the helper:

```
# Simplified version of the factor() helper, as defined in Advanced R
factor <- function(x = character(), levels = unique(x)) {
```

```
  ind <- match(x, levels)
  validate_factor(new_factor(ind, levels))
}
```

To improve the `factor()` helper we choose to return an informative error message instead.

```
factor2 <- function(x, levels = unique(x)) {
  new_levels <- match(x, levels)

  # Error if levels don't include all values
  missing <- unique(setdiff(x, levels))
  if (length(missing) > 0) {
    stop(
      "The following values do not occur in the levels of x: ",
      paste0("'", missing, "'", collapse = ", "), ".",
      call. = FALSE
    )
  }

  validate_factor(new_factor(new_levels, levels))
}

# Test
factor2(c("a", "b", "c"), levels = c("a", "b"))
#> Error: The following values do not occur in the levels of x: 'c'.
```

Q3: Carefully read the source code of `factor()`. What does it do that our constructor does not?

A: The original implementation (`base::factor()`) allows more flexible input for `x`. It coerces `x` to character or replaces it with `character(0)` (in case of `NULL`). It also ensures that the `levels` are unique. This is achieved by setting them via `base::levels<-`, which fails when duplicate values are supplied.

Q4: Factors have an optional "contrasts" attribute. Read the help for `C()`, and briefly describe the purpose of the attribute. What type should it have? Rewrite the `new_factor()` constructor to include this attribute.

A: When factor variables (representing nominal or ordinal information) are used in statistical models, they are typically encoded as dummy variables and by default each level is compared with the first factor level. However, many different encodings ("contrasts") are possible, see `https://en.wikipedia.org/wiki/Contrast_(statistics)`.

Within R's formula interface you can wrap a factor in `stats::C()` and specify the contrast of your choice. Alternatively, you can set the `contrasts` attribute of your factor variable, which accepts matrix input. (See `?contr.helmert` or similar for details.)

The `new_factor()` constructor was given in *Advanced R* as:

```
# new_factor() constructor from Advanced R
new_factor <- function(x = integer(), levels = character()) {
  stopifnot(is.integer(x))
  stopifnot(is.character(levels))

  structure(
    x,
    levels = levels,
    class = "factor"
  )
}
```

Our updated `new_factor()` constructor gets a `contrasts` argument, which accepts a numeric matrix or `NULL` (default).

```
# Updated new_factor() constructor
new_factor <- function(
  x = integer(),
  levels = character(),
  contrasts = NULL
) {
  stopifnot(is.integer(x))
  stopifnot(is.character(levels))

  if (!is.null(constrasts)) {
    stopifnot(is.matrix(contrasts) && is.numeric(contrasts))
  }

  structure(
    x,
    levels = levels,
    class = "factor",
    contrasts = contrasts
  )
}
```

Q5: Read the documentation for `utils::as.roman()`. How would you write a constructor for this class? Does it need a validator? What might a helper do?

A: This function transforms numeric input into Roman numbers. It is built on the integer type, which results in the following constructor.

```
new_roman <- function(x = integer()) {
  stopifnot(is.integer(x))
  structure(x, class = "roman")
}
```

The documentation tells us, that only values between 1 and 3899 are uniquely represented, which we then include in our validation function.

```
validate_roman <- function(x) {
  values <- unclass(x)

  if (any(values < 1 | values > 3899)) {
    stop(
      "Roman numbers must fall between 1 and 3899.",
      call. = FALSE
    )
  }

  x
}
```

For convenience, we allow the user to also pass real values to a helper function.

```
roman <- function(x = integer()) {
  x <- as.integer(x)

  validate_roman(new_roman(x))
}

# Test
roman(c(1, 753, 2019))
#> [1] I       DCCLIII MMXIX
roman(0)
#> Error: Roman numbers must fall between 1 and 3899.
```

13.4 Generics and methods

Q1: Read the source code for `t()` and `t.test()` and confirm that `t.test()` is an S3 generic and not an S3 method. What happens if you create an object with class `test` and call `t()` with it? Why?

```
x <- structure(1:10, class = "test")
t(x)
```

A: We can see that `t.test()` is a generic because it calls `UseMethod()`:

```
t.test
#> function (x, ...)
#> UseMethod("t.test")
#> <bytecode: 0x5567887716f0>
#> <environment: namespace:stats>

# or simply call
ftype(t.test)
#> [1] "S3"      "generic"

# The same holds for t()
t
#> function (x)
#> UseMethod("t")
#> <bytecode: 0x55678beb98e8>
#> <environment: namespace:base>
```

Interestingly, R also provides helpers, which list functions that look like methods, but in fact are not:

```
tools::nonS3methods("stats")
#> [1] "anova.lmlist"         "expand.model.frame"  "fitted.values"
#> [4] "influence.measures"   "lag.plot"            "t.test"
#> [7] "plot.spec.phase"      "plot.spec.coherency"
```

When we create an object with class `test`, `t()` dispatches to the `t.default()` method. This happens, because `UseMethod()` simply searches for functions named `paste0("generic", ".", c(class(x), "default"))`.

```
x <- structure(1:10, class = "test")

t(x)
#>      [,1] [,2] [,3] [,4] [,5] [,6] [,7] [,8] [,9] [,10]
#> [1,]    1    2    3    4    5    6    7    8    9    10
#> attr(,"class")
#> [1] "test"
```

However, in older versions of R (pre R 4.0.0; when *Advanced R* was written) this behaviour was slightly different. Instead of dispatching to the `t.default()` method, the `t.test()` generic was erroneously treated as a method of `t()` which then dispatched to `t.test.default()` or (when defined) to `t.test.test()`.

```
# Output in R version 3.6.2
x <- structure(1:10, class = "test")
t(x)
#>
#>  One Sample t-test
#>
#> data:  x
#> t = 5.7446, df = 9, p-value = 0.0002782
#> alternative hypothesis: true mean is not equal to 0
#> 95 percent confidence interval:
#>  3.334149 7.665851
#> sample estimates:
#> mean of x
#>       5.5

t.test.test <- function(x) "Hi!"
t(x)
#>[1] "Hi!"
```

Q2: What generics does the `table` class have methods for?

A: This is a simple application of `sloop::s3_methods_class()`:

```
s3_methods_class("table")
#> # A tibble: 11 x 4
#>   generic       class visible source
#>   <chr>         <chr> <lgl>   <chr>
#> 1 [             table TRUE    base
#> 2 aperm         table TRUE    base
#> 3 as_tibble     table FALSE   registered S3method
```

```
#>  4 as.data.frame table TRUE    base
#>  5 Axis          table FALSE   registered S3method
#>  6 lines         table FALSE   registered S3method
#>  7 plot          table FALSE   registered S3method
#>  8 points        table FALSE   registered S3method
#>  9 print         table TRUE    base
#> 10 summary       table TRUE    base
#> 11 tail          table FALSE   registered S3method
```

Interestingly, the `table` class has a number of methods designed to help plotting with base graphics.

```
x <- rpois(100, 5)
plot(table(x))
```

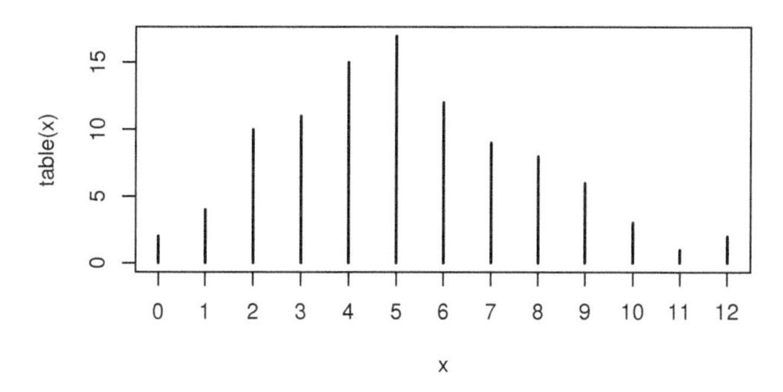

Q3: What generics does the `ecdf` class have methods for?

A: We use the same approach as above:

```
s3_methods_class("ecdf")
#> # A tibble: 4 x 4
#>   generic  class visible source
#>   <chr>    <chr> <lgl>   <chr>
#> 1 plot     ecdf  TRUE    stats
#> 2 print    ecdf  FALSE   registered S3method
#> 3 quantile ecdf  FALSE   registered S3method
#> 4 summary  ecdf  FALSE   registered S3method
```

The methods are primarily designed for display (`plot()`, `print()`, `summary()`), but you can also extract quantiles with `quantile()`.

Q4: Which base generic has the greatest number of defined methods?

A: A little experimentation (and thinking about the most popular functions) suggests that the `print()` generic has the most defined methods.

```
nrow(s3_methods_generic("print"))
#> [1] 260
nrow(s3_methods_generic("summary"))
#> [1] 38
nrow(s3_methods_generic("plot"))
#> [1] 34
```

Let's verify this programmatically with the tools we have learned in this and the previous chapters.

```
library(purrr)

ls(all.names = TRUE, env = baseenv()) %>%
  mget(envir = baseenv()) %>%
  keep(is_function) %>%
  names() %>%
  keep(is_s3_generic) %>%
  map(~ set_names(nrow(s3_methods_generic(.x)), .x)) %>%
  flatten_int() %>%
  sort(decreasing = TRUE) %>%
  head()
#>        print       format            [      summary         plot
#>          260          100           52           38           34
#> as.character
#>           33
```

Q5: Carefully read the documentation for `UseMethod()` and explain why the following code returns the results that it does. What two usual rules of function evaluation does `UseMethod()` violate?

```
g <- function(x) {
  x <- 10
  y <- 10
  UseMethod("g")
}
g.default <- function(x) c(x = x, y = y)

x <- 1
y <- 1
g(x)
#>  x  y
#>  1 10
```

A: Let's take this step by step. If you call `g.default(x)` directly you get `c(1, 1)` as you might expect.

The value bound to `x` comes from the argument, the value from `y` comes from the global environment.

```
g.default(x)
#> x y
#> 1 1
```

But when we call `g(x)` we get `c(1, 10)`:

```
g(x)
#>  x  y
#>  1 10
```

This is seemingly inconsistent: why does `x` come from the value defined inside of `g()`, and `y` still come from the global environment? It's because `UseMethod()` calls `g.default()` in a special way so that variables defined inside the generic are available to methods. The exception are arguments supplied to the function: they are passed on as is and cannot be affected by code inside the generic.

Q6: What are the arguments to `[`? Why is this a hard question to answer?

A: The subsetting operator `[` is a primitive and a generic function, which can be confirmed via `ftype()`.

```
ftype(`[`)
#> [1] "primitive" "generic"
```

For primitive functions `formals([)` returns `NULL` so we need to find another way to determine the functions arguments. One possible way to figure out `[`'s arguments would be to inspect the underlying C source code, which can be searched for via `pryr::show_c_source(.Primitive("["))`.

When we inspect the arguments of some of `[`'s methods, we see that the arguments vary with the class of `x`.

```
names(formals(`[.data.frame`))
#> [1] "x"    "i"    "j"    "drop"
names(formals(`[.table`))
#> [1] "x"    "i"    "j"    "..."  "drop"
names(formals(`[.Date`))
#> [1] "x"    "..."  "drop"
```

```
names(formals(`[.AsIs`))
#> [1] "x"   "i"   "..."
```

To finally get a better overview, we have to put in a little more effort and also use `s3_methods_generic()` again.

```
library(dplyr)

s3_methods_generic("[") %>%
  filter(visible) %>%
  mutate(
    method = paste0("[.", class),
    argnames = purrr::map(method, ~ names(formals(.x))),
    args = purrr::map(method, ~ formals(.x)),
    args = purrr::map2(
      argnames, args,
      ~ paste(.x, .y, sep = " = ")
    ),
    args = purrr::set_names(args, method)
  ) %>%
  pull(args) %>%
  head()
#> $`[.AsIs`
#> [1] "x = "   "i = "   "... = "
#>
#> $`[.data.frame`
#> [1] "x = "
#> [2] "i = "
#> [3] "j = "
#> [4] "drop = if (missing(i)) TRUE else length(cols) == 1"
#>
#> $`[.Date`
#> [1] "x = "           "... = "         "drop = TRUE"
#>
#> $`[.difftime`
#> [1] "x = "           "... = "         "drop = TRUE"
#>
#> $`[.Dlist`
#> [1] "x = "   "i = "   "... = "
#>
#> $`[.DLLInfoList`
#> [1] "x = "   "... = "
```

13.5 Object styles

Q1: Categorise the objects returned by `lm()`, `factor()`, `table()`, `as.Date()`, `as.POSIXct()`, `ecdf()`, `ordered()`, `I()` into the styles described above.

A: We can categorise the return values into the various object styles by observing how the number of observations (https://vctrs.r-lib.org/articles/type-size.html#size) is calculated: For vector style classes, `length(x)` represents the number of observations. Record style objects use a list of equal length elements to represent individual components. For data frames and matrices, the observations are represented by the rows. Scalar style objects use a list to represent a single thing.

This leads us to:

- Vector object-style: `factor()`, `table()`, `as.Date()`, `as.POSIXct()`, `ordered()`
- Record object-style: not observed
- Data frame object-style: not observed
- Scalar object-style: `lm()`, `ecdf()`

The object style of `I()` depends on the input since this function returns a "copy of the object with class `AsIs` prepended to the class(es)".

Q2: What would a constructor function for `lm` objects, `new_lm()`, look like? Use `?lm` and experimentation to figure out the required fields and their types.

A: The constructor needs to populate the attributes of an `lm` object and check their types for correctness. Let's start by creating a simple `lm` object and explore its underlying base type and attributes:

```
mod <- lm(cyl ~ ., data = mtcars)

typeof(mod)
#> [1] "list"

attributes(mod)
#> $names
#>  [1] "coefficients"  "residuals"     "effects"       "rank"
#>  [5] "fitted.values" "assign"        "qr"            "df.residual"
#>  [9] "xlevels"       "call"          "terms"         "model"
#>
#> $class
#> [1] "lm"
```

As mod is built upon a list, we can simply use `map(mod, typeof)` to find out the base types of its elements. (Additionally, we inspect `?lm`, to learn more about the individual attributes.)

```
map_chr(mod, typeof)
#>  coefficients     residuals       effects          rank
#>      "double"      "double"      "double"     "integer"
#> fitted.values        assign            qr   df.residual
#>      "double"     "integer"        "list"     "integer"
#>       xlevels          call         terms         model
#>        "list"    "language"    "language"        "list"
```

Now we should have enough information to write a constructor for new `lm` objects.

```
new_lm <- function(
  coefficients, residuals, effects, rank, fitted.values, assign,
  qr, df.residual, xlevels, call, terms, model
) {

  stopifnot(
    is.double(coefficients), is.double(residuals),
    is.double(effects), is.integer(rank), is.double(fitted.values),
    is.integer(assign), is.list(qr), is.integer(df.residual),
    is.list(xlevels), is.language(call), is.language(terms),
    is.list(model)
  )

  structure(
    list(
      coefficients = coefficients,
      residuals = residuals,
      effects = effects,
      rank = rank,
      fitted.values = fitted.values,
      assign = assign,
      qr = qr,
      df.residual = df.residual,
      xlevels = xlevels,
      call = call,
      terms = terms,
      model = model
    ),
    class = "lm"
```

```
  )
}
```

13.6 Inheritance

Q1: How does `[.Date` support subclasses? How does it fail to support subclasses?

A: `[.Date` calls `.Date` with the result of calling `[` on the parent class, along with `oldClass()`:

```
`[.Date`
#> function (x, ..., drop = TRUE)
#> {
#>     .Date(NextMethod("["), oldClass(x))
#> }
#> <bytecode: 0x55678c510dd8>
#> <environment: namespace:base>
```

`.Date` is kind of like a constructor for date classes, although it doesn't check the input is the correct type:

```
.Date
#> function (xx, cl = "Date")
#> `class<-`(xx, cl)
#> <bytecode: 0x55678ad06508>
#> <environment: namespace:base>
```

`oldClass()` is basically the same as `class()`, except that it doesn't return implicit classes, i.e. it's basically `attr(x, "class")` (looking at the C code that's exactly what it does, except that it also handles S4 objects).

As `oldClass()` is "basically" `class()`, we can rewrite `[.Date` to make the implementation more clear:

```
`[.Date` <- function(x, ..., drop = TRUE) {
  out <- NextMethod("[")
  class(out) <- class(x)
```

```
  out
}
```

So, `[.Date` ensures that the output has the same class as in the input. But what about other attributes that a subclass might possess? They get lost:

```
x <- structure(1:4, test = "test", class = c("myDate", "Date"))
attributes(x[1])
#> $class
#> [1] "myDate" "Date"
```

Q2: R has two classes for representing date time data, `POSIXct` and `POSIXlt`, which both inherit from `POSIXt`. Which generics have different behaviours for the two classes? Which generics share the same behaviour?

A: To answer this question, we have to get the respective generics

```
generics_t  <- s3_methods_class("POSIXt")$generic
generics_ct <- s3_methods_class("POSIXct")$generic
generics_lt <- s3_methods_class("POSIXlt")$generic
```

The generics in `generics_t` with a method for the superclass `POSIXt` potentially share the same behaviour for both subclasses. However, if a generic has a specific method for one of the subclasses, it has to be subtracted:

```
# These generics provide subclass-specific methods
union(generics_ct, generics_lt)
#>  [1] "["              "[["             "[<-"            "as.data.frame"
#>  [5] "as.Date"        "as.list"        "as.POSIXlt"     "c"
#>  [9] "format"         "length<-"       "mean"           "print"
#> [13] "rep"            "split"          "summary"        "Summary"
#> [17] "weighted.mean"  "xtfrm"          "[[<-"           "anyNA"
#> [21] "as.double"      "as.matrix"      "as.POSIXct"     "duplicated"
#> [25] "is.na"          "length"         "names"          "names<-"
#> [29] "sort"           "unique"

# These generics share (inherited) methods for both subclasses
setdiff(generics_t, union(generics_ct, generics_lt))
#>  [1] "-"            "+"            "all.equal"    "as.character"
#>  [5] "Axis"         "cut"          "diff"         "hist"
#>  [9] "is.numeric"   "julian"       "Math"         "months"
#> [13] "Ops"          "pretty"       "quantile"     "quarters"
```

```
#> [17] "round"          "seq"            "str"            "trunc"
#> [21] "weekdays"
```

Q3: What do you expect this code to return? What does it actually return? Why?

```
generic2 <- function(x) UseMethod("generic2")
generic2.a1 <- function(x) "a1"
generic2.a2 <- function(x) "a2"
generic2.b <- function(x) {
  class(x) <- "a1"
  NextMethod()
}

generic2(structure(list(), class = c("b", "a2")))
```

A: When we execute the code above, this is what is happening:

- we pass an object of classes `b` and `a2` to `generic2()`, which prompts R to look for a method`generic2.b()`
- the method `generic2.b()` then changes the class to `a1` and calls `NextMethod()`
- One would think that this will lead R to call `generic2.a1()`, but in fact, as mentioned in *Advanced R*, `NextMethod()`

  > doesn't actually work with the class attribute of the object, but instead uses a special global variable (.Class) to keep track of which method to call next.

  This is why `generic2.a2()` is called instead.

  ```
  generic2(structure(list(), class = c("b", "a2")))
  #> [1] "a2"
  ```

Let's just double check the statement above and evaluate `.Class` explicitly within the `generic2.b()` method.

```
generic2.b <- function(x) {
  class(x) <- "a1"
  print(.Class)
  NextMethod()
}

generic2(structure(list(), class = c("b", "a2")))
```

```
#> [1] "b"  "a2"
#> [1] "a2"
```

13.7 Dispatch details

Q1: Explain the differences in dispatch below:

```
length.integer <- function(x) 10

x1 <- 1:5
class(x1)
#> [1] "integer"
s3_dispatch(length(x1))
#>  * length.integer
#>    length.numeric
#>    length.default
#> => length (internal)

x2 <- structure(x1, class = "integer")
class(x2)
#> [1] "integer"
s3_dispatch(length(x2))
#> => length.integer
#>    length.default
#>  * length (internal)
```

A: `class()` returns `integer` in both cases. However, while the class of `x1` is created implicitly and inherits from the `numeric` class, the class of `x2` is set explicitly. This is important because `length()` is an internal generic and internal generics only dispatch to methods when the class attribute has been set, i.e. internal generics do not use implicit classes.

An object has no explicit class if `attr(x, "class")` returns `NULL`:

```
attr(x1, "class")
#> NULL
attr(x2, "class")
#> [1] "integer"
```

To see the relevant classes for the S3 dispatch, one can use `sloop::s3_class()`:

```
s3_class(x1)  # implicit
#> [1] "integer" "numeric"

s3_class(x2)  # explicit
#> [1] "integer"
```

For a better understanding of `s3_dipatch()`'s output we quote from `?s3_dispatch`:

- => method exists and is found by UseMethod().
- -> method exists and is used by NextMethod().
- * method exists but is not used.
- Nothing (and greyed out in console): method does not exist.

Q2: What classes have a method for the `Math()` group generic in base R? Read the source code. How do the methods work?

A: The following functions belong to this group (see ?`Math`):

- `abs`, `sign`, `sqrt`, `floor`, `ceiling`, `trunc`, `round`, `signif`
- `exp`, `log`, `expm1`, `log1p`, `cos`, `sin`, `tan`, `cospi`, `sinpi`, `tanpi`, `acos`, `asin`, `atan`, `cosh`, `sinh`, `tanh`, `acosh`, `asinh`, `atanh`
- `lgamma`, `gamma`, `digamma`, `trigamma`
- `cumsum`, `cumprod`, `cummax`, `cummin`

The following classes have a method for this group generic:

```
s3_methods_generic("Math")
#> # A tibble: 8 x 4
#>   generic class      visible source
#>   <chr>   <chr>      <lgl>   <chr>
#> 1 Math    data.frame TRUE    base
#> 2 Math    Date       TRUE    base
#> 3 Math    difftime   TRUE    base
#> 4 Math    factor     TRUE    base
#> 5 Math    POSIXt     TRUE    base
#> 6 Math    quosure    FALSE   registered S3method
#> 7 Math    vctrs_sclr FALSE   registered S3method
#> 8 Math    vctrs_vctr FALSE   registered S3method
```

To explain the basic idea, we just overwrite the data frame method:

```
Math.data.frame <- function(x) "hello"
```

Now all functions from the math generic group, will return "hello"

```
abs(mtcars)
#> [1] "hello"
exp(mtcars)
#> [1] "hello"
lgamma(mtcars)
#> [1] "hello"
```

Of course, different functions should perform different calculations. Here .Generic comes into play, which provides us with the calling generic as a string

```
Math.data.frame <- function(x, ...) {
  .Generic
}

abs(mtcars)
#> [1] "abs"
exp(mtcars)
#> [1] "exp"
lgamma(mtcars)
#> [1] "lgamma"

rm(Math.data.frame)
```

The original source code of `Math.data.frame()` is a good example on how to invoke the string returned by `.Generic` into a specific method. `Math.factor()` is a good example of a method, which is simply defined for better error messages.

Q3: `Math.difftime()` is more complicated than I described. Why?

A: `Math.difftime()` also excludes cases apart from `abs`, `sign`, `floor`, `ceiling`, `trunc`, `round` and `signif` and needs to return a fitting error message.

For comparison: `Math.difftime()` as defined in *Advanced R*:

```
Math.difftime <- function(x, ...) {
  new_difftime(NextMethod(), units = attr(x, "units"))
}
rm(Math.difftime)
```

`Math.difftime()` as defined in the `{base}` package:

```
Math.difftime
#> function (x, ...)
#> {
#>     switch(.Generic, abs = , sign = , floor = , ceiling = , trunc = ,
#>         round = , signif = {
#>             units <- attr(x, "units")
#>             .difftime(NextMethod(), units)
#>         }, stop(gettextf("'%s' not defined for \"difftime\" objects",
#>             .Generic), domain = NA))
#> }
#> <bytecode: 0x55678c808148>
#> <environment: namespace:base>
```

14

R6

Prerequisites

To solve the exercises in this chapter we will have to create R6 objects, which are implemented in the `{R6}` package [Chang, 2020].

```
library(R6)
```

14.2 Classes and methods

Q1: Create a bank account R6 class that stores a balance and allows you to deposit and withdraw money. Create a subclass that throws an error if you attempt to go into overdraft. Create another subclass that allows you to go into overdraft, but charges you a fee.

A: Let's start with a basic bank account, similar to the `Accumulator` class in *Advanced R*.

```
BankAccount <- R6Class(
  classname = "BankAccount",
  public = list(
    balance = 0,
    deposit = function(dep = 0) {
      self$balance <- self$balance + dep
      invisible(self)
    },
    withdraw = function(draw) {
      self$balance <- self$balance - draw
      invisible(self)
    }
```

DOI: 10.1201/9781003175414-14

```
  )
)
```

To test this class, we create one instance and leave it with a negative balance.

```
my_account <- BankAccount$new()
my_account$balance
#> [1] 0

my_account$
  deposit(5)$
  withdraw(15)$
  balance
#> [1] -10
```

Now, we create the first subclass that prevents us from going into overdraft and throws an error in case we attempt to withdraw more than our current balance.

```
BankAccountStrict <- R6Class(
  classname = "BankAccountStrict",
  inherit = BankAccount,
  public = list(
    withdraw = function(draw = 0) {
      if (self$balance - draw < 0) {
        stop("Your `withdraw` must be smaller ",
          "than your `balance`.",
          call. = FALSE
        )
      }
      super$withdraw(draw = draw)
    }
  )
)
```

This time our test should throw an error.

```
my_strict_account <- BankAccountStrict$new()
my_strict_account$balance
#> [1] 0

my_strict_account$
```

```
  deposit(5)$
  withdraw(15)
#> Error: Your `withdraw` must be smaller than your `balance`.

my_strict_account$balance
#> [1] 5
```

Finally, we create another subclass that charges a constant fee of 1 for each withdrawal which leaves the account with a negative balance.

```
BankAccountCharging <- R6Class(
  classname = "BankAccountCharging",
  inherit = BankAccount,
  public = list(
    withdraw = function(draw = 0) {
      if (self$balance - draw < 0) {
        draw <- draw + 1
      }
      super$withdraw(draw = draw)
    }
  )
)
```

Let's take a look at the implemented functionality. We expect a final balance of -12, because we pay the fee twice.

```
my_charging_account <- BankAccountCharging$new()
my_charging_account$balance
#> [1] 0

my_charging_account$
  deposit(5)$
  withdraw(15)$
  withdraw(0)

my_charging_account$balance
#> [1] -12
```

Q2: Create an R6 class that represents a shuffled deck of cards. You should be able to draw cards from the deck with `$draw(n)`, and return all cards to the deck and reshuffle with `$reshuffle()`. Use the following code to make a vector of cards.

```
suit <- c("SPADE", "HEARTS", "DIAMOND", "CLUB")
value <- c("A", 2:10, "J", "Q", "K")
cards <- paste(rep(value, 4), suit)
```

(This question was altered slightly to avoid the unicode characters.)

A: Our new `ShuffledDeck` class will use `sample()` and positive integer subsetting to implement the reshuffling and drawing functionality. We also add a check, so you cannot draw more cards than there are left in the deck.

```
ShuffledDeck <- R6Class(
  classname = "ShuffledDeck",
  public = list(
    deck = NULL,
    initialize = function(deck = cards) {
      self$deck <- sample(deck)
    },
    reshuffle = function() {
      self$deck <- sample(cards)
      invisible(self)
    },
    n = function() {
      length(self$deck)
    },
    draw = function(n = 1) {
      if (n > self$n()) {
        stop("Only ", self$n(), " cards remaining.", call. = FALSE)
      }

      output <- self$deck[seq_len(n)]
      self$deck <- self$deck[-seq_len(n)]
      output
    }
  )
)
```

To test this class, we create a deck (initialise an instance), draw all the cards, then reshuffle, checking we get different cards each time.

```
my_deck <- ShuffledDeck$new()

my_deck$draw(52)
#>  [1] "6 SPADE"    "10 DIAMOND" "Q CLUB"     "J SPADE"    "Q HEARTS"
```

```
#>  [6] "8 DIAMOND"  "5 DIAMOND"  "4 CLUB"     "9 CLUB"     "9 SPADE"
#> [11] "5 SPADE"    "3 HEARTS"   "J CLUB"     "2 DIAMOND"  "K SPADE"
#> [16] "2 HEARTS"   "2 SPADE"    "8 SPADE"    "8 HEARTS"   "6 HEARTS"
#> [21] "7 HEARTS"   "6 CLUB"     "K CLUB"     "3 CLUB"     "10 SPADE"
#> [26] "3 DIAMOND"  "Q SPADE"    "9 HEARTS"   "J DIAMOND"  "7 DIAMOND"
#> [31] "9 DIAMOND"  "7 SPADE"    "4 DIAMOND"  "10 HEARTS"  "2 CLUB"
#> [36] "4 SPADE"    "4 HEARTS"   "8 CLUB"     "K HEARTS"   "A SPADE"
#> [41] "A HEARTS"   "5 HEARTS"   "A DIAMOND"  "5 CLUB"     "7 CLUB"
#> [46] "Q DIAMOND"  "A CLUB"     "10 CLUB"    "3 SPADE"    "K DIAMOND"
#> [51] "J HEARTS"   "6 DIAMOND"
my_deck$draw(10)
#> Error: Only 0 cards remaining.

my_deck$reshuffle()$draw(5)
#> [1] "6 DIAMOND" "2 CLUB"    "Q DIAMOND" "9 CLUB"    "J DIAMOND"
my_deck$reshuffle()$draw(5)
#> [1] "8 CLUB"    "9 SPADE"   "2 SPADE"   "Q HEARTS" "6 SPADE"
```

Q3: Why can't you model a bank account or a deck of cards with an S3 class?

A: Because S3 classes obey R's usual semantics of copy-on-modify: every time you deposit money into your bank account or draw a card from the deck, you'd get a new copy of the object.

It is possible to combine S3 classes with an environment (which is how R6 works), but it is ill-advised to create an object that looks like a regular R object but has reference semantics.

Q4: Create an R6 class that allows you to get and set the current time zone. You can access the current time zone with `Sys.timezone()` and set it with `Sys.setenv(TZ = "newtimezone")`. When setting the time zone, make sure the new time zone is in the list provided by `OlsonNames()`.

A: To create an R6 class that allows us to get and set the time zone, we provide the respective functions as public methods to the R6 class.

```
Timezone <- R6Class(
  classname = "Timezone",
  public = list(
    get = function() {
      Sys.timezone()
    },
    set = function(value) {
      stopifnot(value %in% OlsonNames())
```

```
      old <- self$get()
      Sys.setenv(TZ = value)
      invisible(old)
    }
  )
)
```

(When setting, we return the old value invisibly because this makes it easy to restore the previous value.)

Now, let us create one instance of this class and test, if we can set and get the time zone as intended.

```
tz <- Timezone$new()

old <- tz$set("Antarctica/South_Pole")
tz$get()
#> [1] "Europe/Berlin"

tz$set(old)
tz$get()
#> [1] "Europe/Berlin"
```

Q5: Create an R6 class that manages the current working directory. It should have `$get()` and `$set()` methods.

A: Take a look at the following implementation, which is quite minimalistic:

```
WorkingDirectory <- R6Class(
  classname = "WorkingDirectory",
  public = list(
    get = function() {
      getwd()
    },
    set = function(value) {
      setwd(value)
    }
  )
)
```

Q6: Why can't you model the time zone or current working directory with an S3 class?

A: Because S3 classes are not suitable for modelling a state that changes over time. S3 methods should (almost) always return the same result when called with the same inputs.

Q7: What base type are R6 objects built on top of? What attributes do they have?

A: R6 objects are built on top of environments. They have a `class` attribute, which is a character vector containing the class name, the name of any super classes (if existent) and the string `"R6"` as the last element.

14.3 Controlling access

Q1: Create a bank account class that prevents you from directly setting the account balance, but that you can still withdraw from and deposit to. Throw an error if you attempt to go into overdraft.

A: To fulfil this requirement, we make balance a private field. The user has to use the `$deposit()` and `$withdraw()` methods which have access to the balance field.

```
BankAccountStrict2 <- R6Class(
  classname = "BankAccountStrict2",
  public = list(
    deposit = function(dep = 0) {
      private$balance <- private$balance + dep
      invisible(self)
    },
    withdraw = function(draw = 0) {
      if (private$balance - draw < 0) {
        stop(
          "Your `withdraw` must be smaller ",
          "than your `balance`.",
          call. = FALSE
        )
      }
      private$balance <- private$balance - draw
      invisible(self)
    }
  ),
  private = list(
    balance = 0
```

```
  )
)
```

To test our new class, we create an instance and try to go into overdraft.

```
my_account_strict_2 <- BankAccountStrict2$new()

my_account_strict_2$deposit(5)
my_account_strict_2$withdraw(10)
#> Error: Your `withdraw` must be smaller than your `balance`.
```

Q2: Create a class with a write-only `$password` field. It should have `$check_password(password)` method that returns `TRUE` or `FALSE`, but there should be no way to view the complete password.

A: To protect the password from changes and direct access, the password will be a private field. Further, our `Password` will get its own print method which hides the password.

```
Password <- R6Class(
  classname = "Password",
  public = list(
    print = function(...) {
      cat("<Password>: ********\n")
      invisible(self)
    },
    set = function(value) {
      private$password <- value
    },
    check = function(password) {
      identical(password, private$password)
    }
  ),
  private = list(
    password = NULL
  )
)
```

Let's create one instance of our new class and confirm that the password is neither accessible nor visible, but still check-able.

```
my_pw <- Password$new()
my_pw$set("snuffles")
```

```
my_pw$password
#> NULL
my_pw
#> <Password>: ********
my_pw$check("snuggles")
#> [1] FALSE
my_pw$check("snuffles")
#> [1] TRUE
```

Q3: Extend the `Rando` class with another active binding that allows you to access the previous random value. Ensure that active binding is the only way to access the value.

A: To access the previous random value from an instance, we add a private `$last_random` field to our class, and we modify `$random()` to write to this field, whenever it is called. To access the `$last_random` field we provide `$previous()`.

```
Rando <- R6::R6Class(
  classname = "Rando",
  private = list(
    last_random = NULL
  ),
  active = list(
    random = function(value) {
      if (missing(value)) {
        private$last_random <- runif(1)
        private$last_random
      } else {
        stop("Can't set `$random`.", call. = FALSE)
      }
    },
    previous = function(value) {
      if (missing(value)) {
        private$last_random
      }
    }
  )
)
```

Now, we initiate a new `Rando` object and see if it behaves as expected.

```
x <- Rando$new()
x$random
```

```
#> [1] 0.349
x$random
#> [1] 0.947
x$previous
#> [1] 0.947
```

Q4: Can subclasses access private fields/methods from their parent? Perform an experiment to find out.

A: To find out if private fields/methods can be accessed from subclasses, we first create a class `A` with a private field `foo` and a private method `bar()`. Afterwards, an instance of a subclass `B` is created and calls the `foobar()` methods, which tries to access the `foo` field and the `bar()` method from its superclass `A`.

```
A <- R6Class(
  classname = "A",
  private = list(
    field = "foo",
    method = function() {
      "bar"
    }
  )
)

B <- R6Class(
  classname = "B",
  inherit = A,
  public = list(
    test = function() {
      cat("Field:  ", super$field, "\n", sep = "")
      cat("Method: ", super$method(), "\n", sep = "")
    }
  )
)

B$new()$test()
#> Field:
#> Method: bar
```

We conclude that subclasses can access private methods from their superclasses, but not private fields.

14.4 Reference semantics

Q1: Create a class that allows you to write a line to a specified file. You should open a connection to the file in `$initialize()`, append a line using `cat()` in `$append_line()`, and close the connection in `$finalize()`.

A: Our `FileWriter` class will create a connection to a file at initialization. Therefore, we open a connection to a user specified file during the initialisation. Note that we need to set `open = "a"` in `file()` to open connection for appending text. Otherwise, `cat()` would only work when applied to files, but not with connections as explicitly asked for in the exercise. Further, we add the `append_line()` method and a `close()` statement as finalizer.

```
FileWriter <- R6::R6Class(
  classname = "FileWriter",
  public = list(
    con = NULL,
    initialize = function(filename) {
      self$con <- file(filename, open = "a")
    },

    finalize = function() {
      close(self$con)
    },

    append_line = function(x) {
      cat(x, "\n", sep = "", file = self$con)
    }
  )
)
```

Let's see, if new instances of our class work as expected.

```
tmp_file <- tempfile()
my_fw <- FileWriter$new(tmp_file)

readLines(tmp_file)
#> character(0)
my_fw$append_line("First")
my_fw$append_line("Second")
readLines(tmp_file)
#> [1] "First"  "Second"
```

15

S4

Prerequisites

We load the {methods} package [R Core Team, 2020] as it contains the S4 object-oriented programming system.

```
library(methods)
```

15.2 Basics

Q1: `lubridate::period()` returns an S4 class. What slots does it have? What class is each slot? What accessors does it provide?

A: Objects of the S4 `Period` class have six slots named `year`, `month`, `day`, `hour`, `minute`, and `.Data` (which contains the number of seconds). All slots are of type double. Most fields can be retrieved by an identically named accessor (e.g. `lubridate::year()` will return the field), use `second()` to get the `.Data` slot.

As a short example, we create a period of 1 second, 2 minutes, 3 hours, 4 days and 5 weeks.

```
example_12345 <- lubridate::period(
  c(1, 2, 3, 4, 5),
  c("second", "minute", "hour", "day", "week")
)
```

This should add up to a period of 39 days, 3 hours, 2 minutes and 1 second.

DOI: 10.1201/9781003175414-15

```
example_12345
#> [1] "39d 3H 2M 1S"
```

When we inspect example_12345, we see the fields and infer that the seconds are stored in the .Data field.

```
str(example_12345)
#> Formal class 'Period' [package "lubridate"] with 6 slots
#>   ..@ .Data : num 1
#>   ..@ year  : num 0
#>   ..@ month : num 0
#>   ..@ day   : num 39
#>   ..@ hour  : num 3
#>   ..@ minute: num 2
```

Q2: What other ways can you find help for a method? Read ?"?" and summarise the details.

A: Besides adding ? in front of a function call (i.e. ?method()), we may find:

- general documentation for a generic via ?genericName
- general documentation for the methods of a generic via methods?genericName
- documentation for a specific method via ClassName?methodName.

15.3 Classes

Q1: Extend the Person class with fields to match utils::person(). Think about what slots you will need, what class each slot should have, and what you'll need to check in your validity method.

A: The Person class from *Advanced R* contains the slots name and age. The person class from the {utils} package contains the slots given (vector of given names), family, role, email and comment (see ?utils::person).

All slots from utils::person() besides role must be of type character and length 1. The entries in the role slot must match one of the following abbreviations "aut", "com", "cph", "cre", "ctb", "ctr", "dtc", "fnd", "rev", "ths", "trl". Therefore, role might be of different length than the other slots and we'll add a corresponding constraint within the validator.

```
# Definition of the Person class
setClass("Person",
  slots = c(
    age = "numeric",
    given = "character",
    family = "character",
    role = "character",
    email = "character",
    comment = "character"
  ),
  prototype = list(
    age = NA_real_,
    given = NA_character_,
    family = NA_character_,
    role = NA_character_,
    email = NA_character_,
    comment = NA_character_
  )
)

# Helper to create instances of the Person class
Person <- function(given, family,
                   age = NA_real_,
                   role = NA_character_,
                   email = NA_character_,
                   comment = NA_character_) {
  age <- as.double(age)

  new("Person",
    age = age,
    given = given,
    family = family,
    role = role,
    email = email,
    comment = comment
  )
}

# Validator to ensure that each slot is of length one
setValidity("Person", function(object) {
  invalids <- c()
  if (length(object@age)     != 1 ||
      length(object@given)   != 1 ||
      length(object@family)  != 1 ||
```

```
    length(object@email)   != 1 ||
    length(object@comment) != 1) {
  invalids <- paste0("@name, @age, @given, @family, @email, ",
                     "@comment must be of length 1")
}

known_roles <- c(
  NA_character_, "aut", "com", "cph", "cre", "ctb",
  "ctr", "dtc", "fnd", "rev", "ths", "trl"
)

if (!all(object@role %in% known_roles)) {
  paste(
    "@role(s) must be one of",
    paste(known_roles, collapse = ", ")
  )
}

if (length(invalids)) return(invalids)
TRUE
})
#> Class "Person" [in ".GlobalEnv"]
#>
#> Slots:
#>
#> Name:         age      given     family       role      email    comment
#> Class:    numeric character character character character character
```

Q2: What happens if you define a new S4 class that doesn't have any slots? (Hint: read about virtual classes in ?setClass.)

A: It depends on the other arguments. If we inherit from another class, we get the same slots. But something interesting happens if we don't inherit from an existing class. We get a virtual class. A virtual class can't be instantiated:

```
setClass("Human")
new("Human")
#> Error in new("Human"): trying to generate an object from a virtual
#> class ("Human")
```

But can be inherited from:

```
setClass("Programmer", contains = "Human")
```

Q3: Imagine you were going to reimplement factors, dates, and data frames in S4. Sketch out the `setClass()` calls that you would use to define the classes. Think about appropriate `slots` and `prototype`.

A: For all these classes we need one slot for the data and one slot per attribute. Keep in mind, that inheritance matters for ordered factors and dates. For data frames, special checks like equal lengths of the underlying list's elements should be done within a validator.

For simplicity we don't introduce an explicit subclass for ordered factors. Instead, we introduce `ordered` as a slot.

```
setClass("Factor",
  slots = c(
    data = "integer",
    levels = "character",
    ordered = "logical"
  ),
  prototype = list(
    data = integer(),
    levels = character(),
    ordered = FALSE
  )
)
```

```
new("Factor", data = c(1L, 2L), levels = letters[1:3])
#> An object of class "Factor"
#> Slot "data":
#> [1] 1 2
#>
#> Slot "levels":
#> [1] "a" "b" "c"
#>
#> Slot "ordered":
#> [1] FALSE
```

The `Date2` class stores its dates as integers, similarly to base R which uses doubles. Dates don't have any other attributes.

```
setClass("Date2",
  slots = list(
    data = "integer"
  ),
  prototype = list(
```

```
    data = integer()
  )
)

new("Date2", data = 1L)
#> An object of class "Date2"
#> Slot "data":
#> [1] 1
```

Our `DataFrame` class consists of a list and a slot for `row.names`. Most of the logic (e.g. checking that all elements of the list are a vector, and that they all have the same length) would need to be part of a validator.

```
setClass("DataFrame",
  slots = c(
    data = "list",
    row.names = "character"
  ),
  prototype = list(
    data = list(),
    row.names = character(0)
  )
)

new("DataFrame", data = list(a = 1, b = 2))
#> An object of class "DataFrame"
#> Slot "data":
#> $a
#> [1] 1
#>
#> $b
#> [1] 2
#>
#>
#> Slot "row.names":
#> character(0)
```

15.4 Generics and methods

Q1: Add `age()` accessors for the `Person` class.

A: We implement the accessors via an `age()` generic, with a method for the `Person` class and a corresponding replacement function `age<-`:

```
setGeneric("age", function(x) standardGeneric("age"))
#> [1] "age"
setMethod("age", "Person", function(x) x@age)

setGeneric("age<-", function(x, value) standardGeneric("age<-"))
#> [1] "age<-"
setMethod("age<-", "Person", function(x, value) {
  x@age <- value
  validObject(x)
  x
})
```

Q2: In the definition of the generic, why is it necessary to repeat the name of the generic twice?

A: Within `setGeneric()` the `name` (1st argument) is needed as the name of the generic. Then, the name also explicitly incorporates method dispatch via `standardGeneric()` within the generic's body (`def` parameter of `setGeneric()`). This behaviour is similar to `UseMethod()` in S3.

Q3: Why does the `show()` method defined in section 15.4.3 (`https://adv-r.hadley.nz/s4.html#show-method`) use `is(object)[[1]]`? (Hint: try printing the employee subclass.)

A: `is(object)` returns the class of the object. `is(object)` also contains the superclass, for subclasses like `Employee`. In order to always return the most specific class (the subclass), `show()` returns the first element of `is(object)`.

Q4: What happens if you define a method with different argument names to the generic?

A: It depends. We first create the object `hadley` of class `Person`:

```
.Person <- setClass(
  "Person",
  slots = c(name = "character", age = "numeric")
)
```

```
hadley <- .Person(name = "Hadley")
hadley
#> An object of class "Person"
#> Slot "name":
#> [1] "Hadley"
#>
#> Slot "age":
#> numeric(0)
```

Now let's see which arguments can be supplied to the `show()` generic.

```
formals("show")
#> $object
```

Usually, we would use this argument when defining a new method.

```
setMethod("show", "Person", function(object) {
  cat(object@name, "creates hard exercises")
})

hadley
#> Hadley creates hard exercises
```

When we supply another name as a first element of our method (e.g. `x` instead of `object`), this element will be matched to the correct `object` argument and we receive a warning. Our method will work, though:

```
setMethod("show", "Person", function(x) {
  cat(x@name, "creates hard exercises")
})
#> Warning: For function 'show', signature 'Person': argument in method
#> definition changed from (x) to (object)

hadley
#> Hadley creates hard exercises
```

If we add more arguments to our method than our generic can handle, we will get an error.

```
setMethod("show", "Person", function(x, y) {
  cat(x@name, "is", x@age, "years old")
})
```

```
#> Error in conformMethod(signature, mnames, fnames, f, fdef,
#> definition): in method for 'show' with signature 'object="Person"':
#> formal arguments (object = "Person") omitted in the method
#> definition cannot be in the signature
```

If we do this with arguments added to the correctly written `object` argument, we will receive an informative error message. It states that we could add other argument names for generics, which can take the ... argument.

```
setMethod("show", "Person", function(object, y) {
  cat(object@name, "is", object@age, "years old")
})
#> Error in rematchDefinition(definition, fdef, mnames, fnames,
#> signature): methods can add arguments to the generic 'show' only if
#> '...' is an argument to the generic
```

15.5 Method dispatch

Q1: Draw the method graph for `f(😅, 😽)`.

A: Look at the graph and repeat after me: "I will keep my class structure simple and use multiple inheritance sparingly".

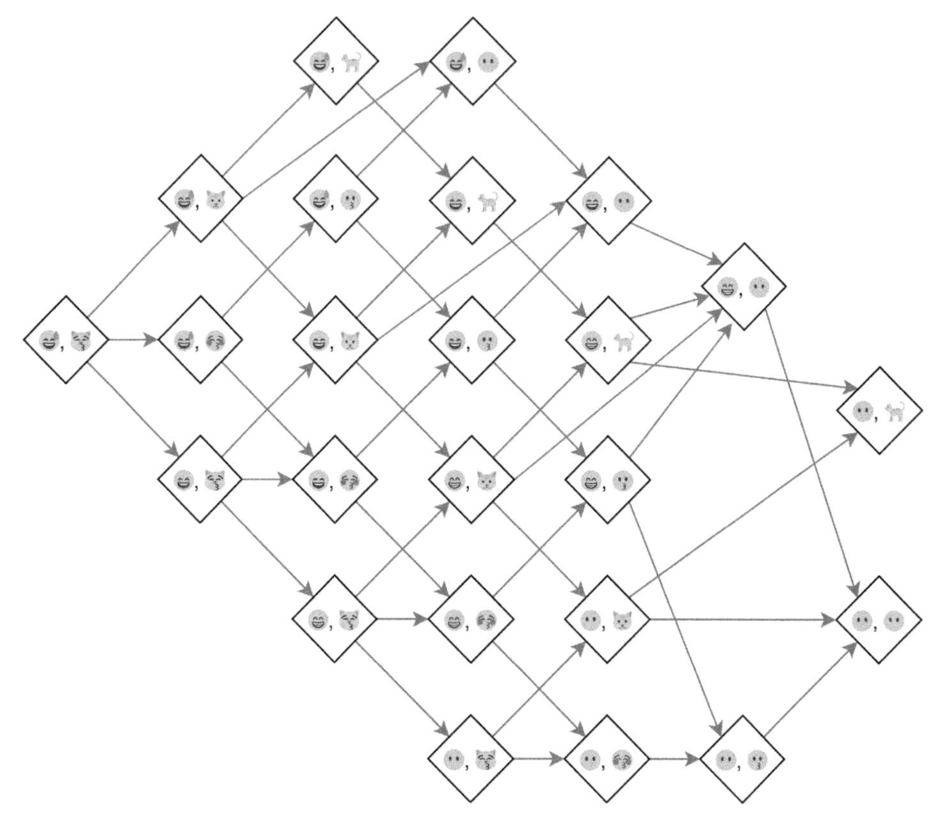

Q2: Draw the method graph for `f(`😃, 😉, 😙`)`.

A: We see that the method graph below looks simpler than the one above. Relatively speaking, multiple dispatch seems to introduce less complexity than multiple inheritance. Use it with care, though!

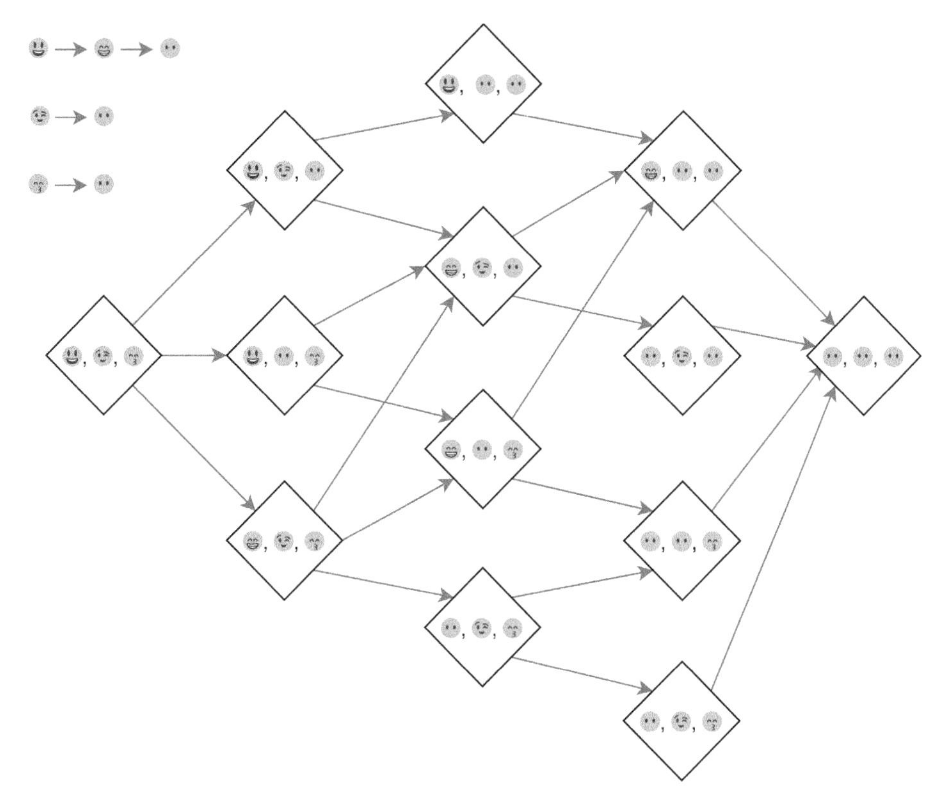

Q3: Take the last example which shows multiple dispatch over two classes that use multiple inheritance. What happens if you define a method for all terminal classes? Why does method dispatch not save us much work here?

A: We will introduce ambiguity, since one class has distance 2 to all terminal nodes and the other four have distance 1 to two terminal nodes each. To resolve this ambiguity we have to define five more methods, one per class combination.

15.6 S4 and S3

Q1: What would a full `setOldClass()` definition look like for an ordered factor (i.e. add `slots` and `prototype` to the definition above)?

A: The purpose of `setOldClass()` lies in registering an S3 class as a "formally defined class", so that it can be used within the S4 object-oriented programming system. When using it, we may provide the argument `S4Class`, which will inherit the slots and their default values (prototype) to the registered class.

Let's build an S4 `OrderedFactor` on top of the S3 factor in such a way.

```
setOldClass("factor")    # use build-in definition for brevity

OrderedFactor <- setClass(
  "OrderedFactor",
  contains = "factor",   # inherit from registered S3 class
  slots = c(
    levels = "character",
    ordered = "logical"  # add logical order slot
  ),
  prototype = structure(
    integer(),
    levels = character(),
    ordered = logical()  # add default value
  )
)
```

We can now register the (S3) ordered-class, while providing an "S4 template". We can also use the S4-class to create new object directly.

```
setOldClass("ordered", S4Class = "OrderedFactor")

x <- OrderedFactor(
  c(1L, 2L, 2L),
  levels = c("a", "b", "c"),
  ordered = TRUE
)
str(x)
#> Formal class 'OrderedFactor' [package ".GlobalEnv"] with 4 slots
#>   ..@ .Data   : int [1:3] 1 2 2
#>   ..@ levels  : chr [1:3] "a" "b" "c"
#>   ..@ ordered : logi TRUE
#>   ..@ .S3Class: chr "factor"
```

Q2: Define a `length` method for the `Person` class.

A: We keep things simple and will just return `"180cm"` when the `length()` method is called on a `Person` object. The method can be defined either as an S3 or S4 method.

```
length.Person <- function(x) "180cm"  # S3
setMethod("length", "Person", function(x) "180cm")  # S4
```

Part IV

Metaprogramming

18

Expressions

Prerequisites

To capture and compute on expressions, and to visualise them, we will load the `{rlang}` [Henry and Wickham, 2020b] and the `{lobstr}` [Wickham, 2019a] packages.

```
library(rlang)
library(lobstr)
```

18.2 Abstract syntax trees

Q1: Reconstruct the code represented by the trees below:

```
#> █─f
#> └─█─g
#>   └─█─h
#> █─`+`
#> ├─█─`+`
#> │ ├─1
#> │ └─2
#> └─3
#> █─`*`
#> ├─█─`(`
#> │ └─█─`+`
#> │   ├─x
#> │   └─y
#> └─z
```

A: Let the source (of the code chunks above) be with you and show you how the ASTs (abstract syntax trees) were produced.

DOI: 10.1201/9781003175414-18

```
ast(f(g(h())))
#> █─f
#> └─█─g
#>   └─█─h

ast(1 + 2 + 3)
#> █─`+`
#> ├─█─`+`
#> │ ├─1
#> │ └─2
#> └─3

ast((x + y) * z)
#> █─`*`
#> ├─█─`(`
#> │ └─█─`+`
#> │   ├─x
#> │   └─y
#> └─z
```

Q2: Draw the following trees by hand then check your answers with `ast()`.

```
f(g(h(i(1, 2, 3))))
f(1, g(2, h(3, i())))
f(g(1, 2), h(3, i(4, 5)))
```

A: Let us delegate the drawing to the `{lobstr}` package.

```
ast(f(g(h(i(1, 2, 3)))))
#> █─f
#> └─█─g
#>   └─█─h
#>     └─█─i
#>       ├─1
#>       ├─2
#>       └─3

ast(f(1, g(2, h(3, i()))))
#> █─f
#> ├─1
#> └─█─g
#>   ├─2
#>   └─█─h
```

```
#>      ├─3
#>      └─█─i

ast(f(g(1, 2), h(3, i(4, 5))))
#> █─f
#> ├─█─g
#> │ ├─1
#> │ └─2
#> └─█─h
#>   ├─3
#>   └─█─i
#>     ├─4
#>     └─5
```

Q3: What's happening with the ASTs below? (Hint: carefully read `?"^"`)

```
ast(`x` + `y`)
#> █─`+`
#> ├─x
#> └─y
ast(x ** y)
#> █─`^`
#> ├─x
#> └─y
ast(1 -> x)
#> █─`<-`
#> ├─x
#> └─1
```

A: ASTs start function calls with the name of the function. This is why the call in the first expression is translated into its prefix form. In the second case, `**` is translated by R's parser into `^`. In the last AST, the expression is flipped when R parses it:

```
str(expr(x ** y))
#>  language x^y
str(expr(a -> b))
#>  language b <- a
```

Q4: What is special about the AST below? (Hint: re-read section 6.2.1 (`https://adv-r.hadley.nz/functions.html#fun-components`))

```
ast(function(x = 1, y = 2) {})
#> █─`function`
#> ├─█─x = 1
#> │ └─y = 2
#> ├─█─`{`
#> └─<inline srcref>
```

A: The last leaf of the AST is not explicitly specified in the expression. Instead, the `srcref` attribute, which points to the functions source code, is automatically created by base R.

Q5: What does the call tree of an `if` statement with multiple `else if` conditions look like? Why?

A: The AST of nested `else if` statements might look a bit confusing because it contains multiple curly braces. However, we can see that in the `else` part of the AST just another expression is being evaluated, which happens to be an `if` statement and so forth.

```
ast(
  if (FALSE) {
    1
  } else if (FALSE) {
    2
  } else if (TRUE) {
    3
  }
)
#> █─`if`
#> ├─FALSE
#> ├─█─`{`
#> │ └─1
#> └─█─`if`
#>   ├─FALSE
#>   ├─█─`{`
#>   │ └─2
#>   └─█─`if`
#>     ├─TRUE
#>     └─█─`{`
#>       └─3
```

We can see the structure more clearly if we avoid the curly braces:

```r
ast(
  if (FALSE) 1
  else if (FALSE) 2
  else if (TRUE) 3
)
#> █─`if`
#> ├─FALSE
#> ├─1
#> └─█─`if`
#>   ├─FALSE
#>   ├─2
#>   └─█─`if`
#>     ├─TRUE
#>     └─3
```

18.3 Expressions

Q1: Which two of the six types of atomic vector can't appear in an expression? Why? Similarly, why can't you create an expression that contains an atomic vector of length greater than one?

A: There is no way to create raws and complex atomics without using a function call (this is only possible for imaginary scalars like `i`, `5i` etc.). But expressions that include a function are *calls*. Therefore, both of these vector types cannot appear in an expression.

Similarly, it is not possible to create an expression that evaluates to an atomic of length greater than one without using a function (e.g. `c()`).

Let's make this observation concrete via an example:

```r
# Atomic
is_atomic(expr(1))
#> [1] TRUE

# Not an atomic (although it would evaluate to an atomic)
is_atomic(expr(c(1, 1)))
#> [1] FALSE
is_call(expr(c(1, 1)))
#> [1] TRUE
```

Q2: What happens when you subset a call object to remove the first element, e.g. `expr(read.csv("foo.csv", header = TRUE))[-1]`. Why?

A: When the first element of a call object is removed, the second element moves to the first position, which is the function to call. Therefore, we get `"foo.csv"(header = TRUE)`.

Q3: Describe the differences between the following call objects.

```
x <- 1:10

call2(median, x, na.rm = TRUE)
call2(expr(median), x, na.rm = TRUE)
call2(median, expr(x), na.rm = TRUE)
call2(expr(median), expr(x), na.rm = TRUE)
```

A: The call objects differ in their first two elements, which are in some cases evaluated before the call is constructed. In the first one, both `median()` and `x` are evaluated and inlined into the call. Therefore, we can see in the constructed call that median is a generic and the `x` argument is 1:10.

```
call2(median, x, na.rm = TRUE)
#> (function (x, na.rm = FALSE, ...)
#> UseMethod("median"))(1:10, na.rm = TRUE)
```

In the following calls we remain with differing combinations. Once, only `x` and once only `median()` gets evaluated.

```
call2(expr(median), x, na.rm = TRUE)
#> median(1:10, na.rm = TRUE)
call2(median, expr(x), na.rm = TRUE)
#> (function (x, na.rm = FALSE, ...)
#> UseMethod("median"))(x, na.rm = TRUE)
```

In the final call neither `x` nor `median()` is evaluated.

```
call2(expr(median), expr(x), na.rm = TRUE)
#> median(x, na.rm = TRUE)
```

Note that all these calls will generate the same result when evaluated. The key difference is when the values bound to the `x` and `median` symbols are found.

Q4: `rlang::call_standardise()` doesn't work so well for the following calls. Why? What makes `mean()` special?

```
call_standardise(quote(mean(1:10, na.rm = TRUE)))
#> mean(x = 1:10, na.rm = TRUE)
call_standardise(quote(mean(n = T, 1:10)))
#> mean(x = 1:10, n = T)
call_standardise(quote(mean(x = 1:10, , TRUE)))
#> mean(x = 1:10, , TRUE)
```

A: The reason for this unexpected behaviour is that `mean()` uses the ... argument and therefore cannot standardise the regarding arguments. Since `mean()` uses S3 dispatch (i.e. `UseMethod()`) and the underlying `mean.default()` method specifies some more arguments, `call_standardise()` can do much better with a specific S3 method.

```
call_standardise(quote(mean.default(1:10, na.rm = TRUE)))
#> mean.default(x = 1:10, na.rm = TRUE)
call_standardise(quote(mean.default(n = T, 1:10)))
#> mean.default(x = 1:10, na.rm = T)
call_standardise(quote(mean.default(x = 1:10, , TRUE)))
#> mean.default(x = 1:10, na.rm = TRUE)
```

Q5: Why does this code not make sense?

```
x <- expr(foo(x = 1))
names(x) <- c("x", "")
```

A: As stated in *Advanced R*

> The first element of a call is always the function that gets called.

Let's see what happens when we run the code

```
x <- expr(foo(x = 1))
x
#> foo(x = 1)

names(x) <- c("x", "")
x
#> foo(1)

names(x) <- c("", "x")
x
#> foo(x = 1)
```

So, giving the first element a name just adds metadata that R ignores.

Q6: Construct the expression `if(x > 1) "a" else "b"` using multiple calls to `call2()`. How does the code structure reflect the structure of the AST?

A: Similar to the prefix version we get

```
call2("if", call2(">", sym("x"), 1), "a", "b")
#> if (x > 1) "a" else "b"
```

When we read the AST from left to right, we get the same structure: Function to evaluate, expression, which is another function and is evaluated first, and two constants which will be evaluated next.

```
ast(`if`(x > 1, "a", "b"))
#> █─`if`
#> ├─█─`>`
#> │ ├─x
#> │ └─1
#> ├─"a"
#> └─"b"
```

18.4 Parsing and grammar

Q1: R uses parentheses in two slightly different ways as illustrated by these two calls:

```
f((1))
`(`(1 + 1)
```

Compare and contrast the two uses by referencing the AST.

A: The trick with these examples lies in the fact that `(` can be a part of R's general prefix function syntax but can also represent a call to the `(` function.

So, in the AST of the first example, we will not see the outer `(` since it is prefix function syntax and belongs to `f()`. In contrast, the inner `(` is a function (represented as a symbol in the AST):

```
ast(f((1)))
#> █─f
```

```
#> └─█─`(`
#>   └─1
```

In the second example, we can see that the outer `(` is a function and the inner `(` belongs to its syntax:

```
ast(`(`(1 + 1))
#> █─`(`
#> └─█─`+`
#>   ├─1
#>   └─1
```

For the sake of clarity, let's also create a third example, where none of the `(` is part of another function's syntax:

```
ast(((1 + 1)))
#> █─`(`
#> └─█─`(`
#>   └─█─`+`
#>     ├─1
#>     └─1
```

Q2: `=` can also be used in two ways. Construct a simple example that shows both uses.

A: `=` is used both for assignment, and for naming arguments in function calls:

```
b = c(c = 1)
```

So, when we play with `ast()`, we can directly see that the following is not possible:

```
ast(b = c(c = 1))
#> Error in ast(b = c(c = 1)): unused argument (b = c(c = 1))
```

We get an error because `b =` makes R looking for an argument called `b`. Since `x` is the only argument of `ast()`, we get an error.

The easiest way around this problem is to wrap this line in `{}`.

```
ast({b = c(c = 1)})
#> █─`{`
#> └─█─`=`
```

```
#>   ├─b
#>   └─█─c
#>     └─c = 1
```

When we ignore the braces and compare the trees, we can see that the first = is used for assignment and the second = is part of the syntax of function calls.

Q3: Does `-2^2` yield 4 or -4? Why?

A: It yields `-4`, because `^` has a higher operator precedence than `-`, which we can verify by looking at the AST (or looking it up under `?"Syntax"`):

```
-2^2
#> [1] -4

ast(-2^2)
#> █─`-`
#> └─█─`^`
#>   ├─2
#>   └─2
```

Q4: What does `!1 + !1` return? Why?

A: The answer is a little surprising:

```
!1 + !1
#> [1] FALSE
```

To answer the "why?", we take a look at the AST:

```
ast(!1 + !1)
#> █─`!`
#> └─█─`+`
#>   ├─1
#>   └─█─`!`
#>     └─1
```

The right `!1` is evaluated first. It evaluates to `FALSE`, because R coerces every non 0 numeric to `TRUE`, when a logical operator is applied. The negation of `TRUE` then equals `FALSE`.

Next `1 + FALSE` is evaluated to `1`, since `FALSE` is coerced to `0`.

Finally `!1` is evaluated to `FALSE`.

Note that if ! had a higher precedence, the intermediate result would be `FALSE + FALSE`, which would evaluate to `0`.

Q5: Why does `x1 <- x2 <- x3 <- 0` work? Describe the two reasons.

A: One reason is that `<-` is right-associative, i.e. evaluation takes place from right to left:

```
x1 <- (x2 <- (x3 <- 0))
```

The other reason is that `<-` invisibly returns the value on the right-hand side.

```
(x3 <- 0)
#> [1] 0
```

Q6: Compare the ASTs of `x + y %+% z` and `x ^ y %+% z`. What have you learned about the precedence of custom infix functions?

A: Let's take a look at the syntax trees:

```
ast(x + y %+% z)
#> █─`+`
#> ├─x
#> └─█─`%+%`
#>   ├─y
#>   └─z
```

Here `y %+% z` will be calculated first and the result will be added to `x`.

```
ast(x ^ y %+% z)
#> █─`%+%`
#> ├─█─`^`
#> │ ├─x
#> │ └─y
#> └─z
```

Here `x ^ y` will be calculated first, and the result will be used as first argument to `%+%()`.

We can conclude that custom infix functions have precedence between addition and exponentiation.

The exact precedence of infix functions can be looked up under `?"Syntax"` where we see that it lies directly behind the sequence operator (`:`) and in front of the multiplication and division operators (`*` and `/`).

Q7: What happens if you call `parse_expr()` with a string that generates multiple expressions, e.g. `parse_expr("x + 1; y + 1")`?

A: In this case `parse_expr()` notices that more than one expression would have to be generated and throws an error.

```
parse_expr("x + 1; y + 1")
#> Error: More than one expression parsed
```

Q8: What happens if you attempt to parse an invalid expression, e.g. `"a +"` or `"f())"`?

A: Invalid expressions will lead to an error in the underlying `parse()` function.

```
parse_expr("a +")
#> Error in parse(text = elt): <text>:2:0: unexpected end of input
#> 1: a +
#>    ^
parse_expr("f())")
#> Error in parse(text = elt): <text>:1:4: unexpected ')'
#> 1: f())
#>        ^
```

```
parse(text = "a +")
#> Error in parse(text = "a +"): <text>:2:0: unexpected end of input
#> 1: a +
#>    ^
parse(text = "f())")
#> Error in parse(text = "f())"): <text>:1:4: unexpected ')'
#> 1: f())
#>        ^
```

Q9: `deparse()` produces vectors when the input is long. For example, the following call produces a vector of length two:

```
expr <- expr(g(a + b + c + d + e + f + g + h + i + j + k + l + m +
                n + o + p + q + r + s + t + u + v + w + x + y + z))

deparse(expr)
```

What does `expr_text()` do instead?

A: `expr_text()` will paste the results from `deparse(expr)` together and use a linebreak (`\n`) as separator.

```
expr <- expr(g(a + b + c + d + e + f + g + h + i + j + k + l + m +
                 n + o + p + q + r + s + t + u + v + w + x + y + z))
deparse(expr)
#> [1] "g(a + b + c + d + e + f + g + h + i + j + k + l + m + n + "
#> [2] "o + p + q + r + s + t + u + v + w + x + y + z)"
expr_text(expr)
#> [1] "g(a + b + c + d + e + f + g + h + i + j + k + l + m + n
#> + \n    o + p + q + r + s + t + u + v + w + x + y + z)"
```

Q10: `pairwise.t.test()` assumes that `deparse()` always returns a length one character vector. Can you construct an input that violates this expectation? What happens?

A: The function `pairwise.t.test()` captures its data arguments (`x` and `g`) so it can print the input expressions along the computed p-values. Prior to R 4.0.0 this used to be implemented via `deparse(substitute(x))` in combination with `paste()`. This could lead to unexpected output, if one of the inputs exceeded the default `width.cutoff` value of 60 characters within `deparse()`. In this case, the expression would be split into a character vector of length greater 1.

```
# Output in R version 3.6.2
d <- 1
pairwise.t.test(2, d + d + d + d + d + d + d + d +
                  d + d + d + d + d + d + d + d + d)
#>  Pairwise comparisons using t tests with pooled SD
#>
#> data:  2 and d + d + d + d + d + d + d + d + d + d + d + d + d + d
#> + d + d +  2 and     d
#>
#> <0 x 0 matrix>
#>
#> P value adjustment method: holm
```

In R 4.0.0 (`https://cran.r-project.org/doc/manuals/r-release/NEWS.html`) `pairwise.t.test()` was updated to use the newly introduced `deparse1()`, which serves as a wrapper around `deparse()`.

> deparse1() is a simple utility added in R 4.0.0 to ensure a string result (character vector of length one), typically used in name construction, as deparse1(substitute(.)).

```
# Output since R 4.0.0
d <- 1
pairwise.t.test(2, d + d + d + d + d + d + d + d +
```

```
                d + d + d + d + d + d + d + d + d)
#>  Pairwise comparisons using t tests with pooled SD
#>
#> data:  2 and d + d + d + d + d + d + d + d + d + d + d + d + d + d
#> + d + d + d
#>
#> <0 x 0 matrix>
#>
#> P value adjustment method: holm
```

18.5 Walking AST with recursive functions

Q1: `logical_abbr()` returns `TRUE` for `T(1, 2, 3)`. How could you modify `logical_abbr_rec()` so that it ignores function calls that use `T` or `F`?

A: We can apply a similar logic as in the assignment example (https://adv-r.hadley.nz/expressions.html#finding-all-variables-created-by-assignment) from *Advanced R*. We just treat it as a special case handled within a sub function called `find_T_call()`, which finds `T()` calls and "bounces them out". Therefore, we also repeat the `expr_type()` helper which tells us if we are in the base or in the recursive case.

```r
expr_type <- function(x) {
  if (rlang::is_syntactic_literal(x)) {
    "constant"
  } else if (is.symbol(x)) {
    "symbol"
  } else if (is.call(x)) {
    "call"
  } else if (is.pairlist(x)) {
    "pairlist"
  } else {
    typeof(x)
  }
}

switch_expr <- function(x, ...) {
  switch(expr_type(x),
         ...,
```

```
        stop("Don't know how to handle type ",
             typeof(x), call. = FALSE))
}


find_T_call <- function(x) {
  if (is_call(x, "T")) {
    x <- as.list(x)[-1]
    purrr::some(x, logical_abbr_rec)
  } else {
    purrr::some(x, logical_abbr_rec)
  }
}

logical_abbr_rec <- function(x) {
  switch_expr(
    x,
    # Base cases
    constant = FALSE,
    symbol = as_string(x) %in% c("F", "T"),

    # Recursive cases
    pairlist = purrr::some(x, logical_abbr_rec),
    call = find_T_call(x)
  )
}

logical_abbr <- function(x) {
  logical_abbr_rec(enexpr(x))
}
```

Now let's test our new `logical_abbr()` function:

```
logical_abbr(T(1, 2, 3))
#> [1] FALSE
logical_abbr(T(T, T(3, 4)))
#> [1] TRUE
logical_abbr(T(T))
#> [1] TRUE
logical_abbr(T())
#> [1] FALSE
logical_abbr()
#> [1] FALSE
```

```
logical_abbr(c(T, T, T))
#> [1] TRUE
```

Q2: `logical_abbr()` works with expressions. It currently fails when you give it a function. Why? How could you modify `logical_abbr()` to make it work? What components of a function will you need to recurse over?

```
f <- function(x = TRUE) {
  g(x + T)
}
```

A: The function currently fails, because `"closure"` is not handled in `switch_expr()` within `logical_abbr_rec()`.

```
logical_abbr(!!f)
#> Error: Don't know how to handle type closure
```

If we want to make it work, we have to write a function to also iterate over the formals and the body of the input function.

Q3: Modify `find_assign` to also detect assignment using replacement functions, i.e. `names(x) <- y`.

A: Let's see what the AST of such an assignment looks like:

```
ast(names(x) <- x)
#> █─`<-`
#> ├─█─names
#> │ └─x
#> └─x
```

So, we need to catch the case where the first two elements are both calls. Further the first call is identical to `<-` and we must return only the second call to see which objects got new values assigned.

This is why we add the following block within another `else` statement in `find_assign_call()`:

```
if (is_call(x, "<-") && is_call(x[[2]])) {
  lhs <- expr_text(x[[2]])
  children <- as.list(x)[-1]
}
```

Let us finish with the whole code, followed by some tests for our new function:

```
flat_map_chr <- function(.x, .f, ...) {
  purrr::flatten_chr(purrr::map(.x, .f, ...))
}

find_assign <- function(x) unique(find_assign_rec(enexpr(x)))

find_assign_call <- function(x) {
  if (is_call(x, "<-") && is_symbol(x[[2]])) {
    lhs <- as_string(x[[2]])
    children <- as.list(x)[-1]
  } else {
    if (is_call(x, "<-") && is_call(x[[2]])) {
      lhs <- expr_text(x[[2]])
      children <- as.list(x)[-1]
    } else {
      lhs <- character()
      children <- as.list(x)
    }}

  c(lhs, flat_map_chr(children, find_assign_rec))
}

find_assign_rec <- function(x) {
  switch_expr(
    x,
    # Base cases
    constant = ,symbol = character(),
    # Recursive cases
    pairlist = flat_map_chr(x, find_assign_rec),
    call = find_assign_call(x)
  )
}

# Tests functionality
find_assign(x <- y)
#> [1] "x"
find_assign(names(x))
#> character(0)
find_assign(names(x) <- y)
#> [1] "names(x)"
find_assign(names(x(y)) <- y)
#> [1] "names(x(y))"
find_assign(names(x(y)) <- y <- z)
#> [1] "names(x(y))" "y"
```

Q4: Write a function that extracts all calls to a specified function.

A: Here we need to delete the previously added else statement and check for a call (not necessarily `<-`) within the first `if()` in `find_assign_call()`. We save a call when we found one and return it later as part of our character output. Everything else stays the same:

```
find_assign_call <- function(x) {
  if (is_call(x)) {
    lhs <- expr_text(x)
    children <- as.list(x)[-1]
  } else {
    lhs <- character()
    children <- as.list(x)
  }

  c(lhs, flat_map_chr(children, find_assign_rec))
}

find_assign_rec <- function(x) {
  switch_expr(
    x,
    # Base cases
    constant = ,
    symbol = character(),

    # Recursive cases
    pairlist = flat_map_chr(x, find_assign_rec),
    call = find_assign_call(x)
  )
}

find_assign(x <- y)
#> [1] "x <- y"
find_assign(names(x(y)) <- y <- z)
#> [1] "names(x(y)) <- y <- z" "names(x(y))"
#> [3] "x(y)"                  "y <- z"
find_assign(mean(sum(1:3)))
#> [1] "mean(sum(1:3))" "sum(1:3)"       "1:3"
```

19

Quasiquotation

Prerequisites

To continue computing on the language, we keep using the `{rlang}` package in this chapter.

```
library(rlang)
```

19.2 Motivation

Q1: For each function in the following base R code, identify which arguments are quoted and which are evaluated.

```
library(MASS)

mtcars2 <- subset(mtcars, cyl == 4)

with(mtcars2, sum(vs))
sum(mtcars2$am)

rm(mtcars2)
```

A: For each argument we first follow the advice from *Advanced R* and execute the argument outside of the respective function. Since `MASS`, `cyl`, `vs` and `am` are not objects contained in the global environment, their execution raises an "Object not found" error. This way we confirm that the respective function arguments are quoted. For the other arguments, we may inspect the source code (and the documentation) to check if any quoting mechanisms are applied or the arguments are evaluated.

DOI: 10.1201/9781003175414-19

```
library(MASS)  # MASS -> quoted
```

library() also accepts character vectors and doesn't quote when character.only is set to TRUE, so library(MASS, character.only = TRUE) would raise an error.

```
mtcars2 <- subset(mtcars, cyl == 4)  # mtcars -> evaluated
# cyl -> quoted

with(mtcars2, sum(vs))  # mtcars2 -> evaluated
# sum(vs) -> quoted

sum(mtcars2$am)  # matcars$am -> evaluated
# am -> quoted by $()
```

When we inspect the source code of rm(), we notice that rm() catches its ... argument as an unevaluated call (in this case a pairlist) via match.call(). This call is then converted into a string for further evaluation.

```
rm(mtcars2)  # mtcars2 -> quoted
```

Q2: For each function in the following tidyverse code, identify which arguments are quoted and which are evaluated.

```
library(dplyr)
library(ggplot2)

by_cyl <- mtcars %>%
  group_by(cyl) %>%
  summarise(mean = mean(mpg))

ggplot(by_cyl, aes(cyl, mean)) + geom_point()
```

A: From the previous exercise we've already learned that library() quotes its first argument.

```
library(dplyr)    # dplyr   -> quoted
library(ggplot2)  # ggplot2 -> quoted
```

In similar fashion, it becomes clear that cyl is quoted by group_by().

```
by_cyl <- mtcars %>%              # mtcars -> evaluated
  group_by(cyl) %>%               # cyl -> quoted
  summarise(mean = mean(mpg))  # mean = mean(mpg) -> quoted
```

To find out what happens in summarise(), we inspect the source code. Tracing down the S3-dispatch of summarise(), we see that the ... argument is quoted in dplyr:::summarise_cols() which is called in the underlying summarise.data.frame() method.

```
dplyr::summarise
#> function (.data, ..., .groups = NULL)
#> {
#>     UseMethod("summarise")
#> }
#> <bytecode: 0x562ef4376378>
#> <environment: namespace:dplyr>
```

```
dplyr:::summarise.data.frame
#> function (.data, ..., .groups = NULL)
#> {
#>     cols <- summarise_cols(.data, ...)
#>     out <- summarise_build(.data, cols)
#>     if (identical(.groups, "rowwise")) {
#>         out <- rowwise_df(out, character())
#>     }
#>     out
#> }
#> <bytecode: 0x562ef469a040>
#> <environment: namespace:dplyr>
```

```
dplyr:::summarise_cols
#> function (.data, ...)
#> {
#>     mask <- DataMask$new(.data, caller_env())
#>     dots <- enquos(...)
#>     dots_names <- names(dots)
#>     auto_named_dots <- names(enquos(..., .named = TRUE))
#>     cols <- list()
#>     sizes <- 1L
#>     chunks <- vector("list", length(dots))
#>     types <- vector("list", length(dots))
```

```
#>
#>     ## function definition abbreviated for clarity ##
#> }
#> <bytecode: 0x55b540c07ca0>
#> <environment: namespace:dplyr>
```

In the following {ggplot2} expression the cyl- and mean-objects are quoted.

```
ggplot(by_cyl,             # by_cyl -> evaluated
       aes(cyl, mean)) +   # aes() -> evaluated
  # cyl, mean -> quoted (via aes)
  geom_point()
```

We can confirm this also by inspecting aes()'s source code.

```
ggplot2::aes
#> function (x, y, ...)
#> {
#>     exprs <- enquos(x = x, y = y, ..., .ignore_empty = "all")
#>     aes <- new_aes(exprs, env = parent.frame())
#>     rename_aes(aes)
#> }
#> <bytecode: 0x562ef4b5e498>
#> <environment: namespace:ggplot2>
```

19.3 Quoting

Q1: How is expr() implemented? Look at its source code.

A: expr() acts as a simple wrapper, which passes its argument to enexpr().

```
expr
#> function (expr)
#> {
#>     enexpr(expr)
#> }
#> <bytecode: 0x562ef4cedaa0>
#> <environment: namespace:rlang>
```

Q2: Compare and contrast the following two functions. Can you predict the output before running them?

```
f1 <- function(x, y) {
  exprs(x = x, y = y)
}
f2 <- function(x, y) {
  enexprs(x = x, y = y)
}
f1(a + b, c + d)
f2(a + b, c + d)
```

A: Both functions are able to capture multiple arguments and will return a named list of expressions. `f1()` will return the arguments defined within the body of `f1()`. This happens because `exprs()` captures the expressions as specified by the developer during the definition of `f1()`.

```
f1(a + b, c + d)
#> $x
#> x
#>
#> $y
#> y
```

`f2()` will return the arguments supplied to `f2()` as specified by the user when the function is called.

```
f2(a + b, c + d)
#> $x
#> a + b
#>
#> $y
#> c + d
```

Q3: What happens if you try to use `enexpr()` with an expression (i.e. `enexpr(x + y)`)? What happens if `enexpr()` is passed a missing argument?

A: In the first case an error is thrown:

```
on_expr <- function(x) {enexpr(expr(x))}
on_expr(x + y)
#> Error: `arg` must be a symbol
```

In the second case a missing argument is returned:

```
on_missing <- function(x) {enexpr(x)}
on_missing()
is_missing(on_missing())
#> [1] TRUE
```

Q4: How are `exprs(a)` and `exprs(a = )` different? Think about both the input and the output.

A: In `exprs(a)` the input `a` is interpreted as a symbol for an unnamed argument. Consequently, the output shows an unnamed list with the first element containing the symbol `a`.

```
out1 <- exprs(a)
str(out1)
#> List of 1
#>  $ : symbol a
```

In `exprs(a = )` the first argument is named `a`, but then no value is provided. This leads to the output of a named list with the first element named `a`, which contains the missing argument.

```
out2 <- exprs(a = )
str(out2)
#> List of 1
#>  $ a: symbol
is_missing(out2$a)
#> [1] TRUE
```

Q5: What are other differences between `exprs()` and `alist()`? Read the documentation for the named arguments of `exprs()` to find out.

A: `exprs()` provides the additional arguments `.named` (= `FALSE`), `.ignore_empty` (`c("trailing", "none", "all")`) and `.unquote_names` (`TRUE`). `.named` allows to ensure that all dots are named. `ignore_empty` allows to specify how empty arguments should be handled for dots (`"trailing"`) or all arguments (`"none"` and `"all"`). Further via `.unquote_names` one can specify if `:=` should be treated like `=`. `:=` can be useful as it supports unquoting (`!!`) on the left-hand side.

Q6: The documentation for `substitute()` says:

> Substitution takes place by examining each component of the parse tree as follows:
>
> - If it is not a bound symbol in `env`, it is unchanged.
> - If it is a promise object (i.e. a formal argument to a function) the expression slot of the promise replaces the symbol.

- If it is an ordinary variable, its value is substituted, unless env is .GlobalEnv in which case the symbol is left unchanged.

Create examples that illustrate each of the above cases.

A: Let's create a new environment `my_env`, which contains no objects. In this case `substitute()` will just return its first argument (`expr`):

```
my_env <- env()
substitute(x, my_env)
#> x
```

When we create a function containing an argument, which is directly returned after substitution, this function just returns the provided expression:

```
foo <- function(x) substitute(x)

foo(x + y * sin(0))
#> x + y * sin(0)
```

In case `substitute()` can find (parts of) the expression in `env`, it will literally substitute. However, unless `env` is `.GlobalEnv`.

```
my_env$x <- 7
substitute(x, my_env)
#> [1] 7

x <- 7
substitute(x, .GlobalEnv)
#> x
```

19.4 Unquoting

Q1: Given the following components:

```
xy <- expr(x + y)
xz <- expr(x + z)
yz <- expr(y + z)
abc <- exprs(a, b, c)
```

Use quasiquotation to construct the following calls:

```
(x + y) / (y + z)                  # (1)
-(x + z) ^ (y + z)                 # (2)
(x + y) + (y + z) - (x + y)        # (3)
atan2(x + y, y + z)                # (4)
sum(x + y, x + y, y + z)           # (5)
sum(a, b, c)                       # (6)
mean(c(a, b, c), na.rm = TRUE)     # (7)
foo(a = x + y, b = y + z)          # (8)
```

A: We combine and unquote the given quoted expressions to construct the desired calls like this:

```
expr(!!xy / !!yz)                    # (1)
#> (x + y)/(y + z)

expr(-(!!xz)^(!!yz))                 # (2)
#> -(x + z)^(y + z)

expr(((!!xy)) + !!yz-!!xy)           # (3)
#> (x + y) + (y + z) - (x + y)

expr(atan2(!!xy, !!yz))              # (4)
#> atan2(x + y, y + z)

expr(sum(!!xy, !!xy, !!yz))          # (5)
#> sum(x + y, x + y, y + z)

expr(sum(!!!abc))                    # (6)
#> sum(a, b, c)

expr(mean(c(!!!abc), na.rm = TRUE))  # (7)
#> mean(c(a, b, c), na.rm = TRUE)

expr(foo(a = !!xy, b = !!yz))        # (8)
#> foo(a = x + y, b = y + z)
```

Q2: The following two calls print the same, but are actually different:

```
(a <- expr(mean(1:10)))
#> mean(1:10)
(b <- expr(mean(!!(1:10))))
```

```
#> mean(1:10)
identical(a, b)
#> [1] FALSE
```

What's the difference? Which one is more natural?

A: It's easiest to see the difference with `lobstr::ast()`:

```
lobstr::ast(mean(1:10))
#> █─mean
#> └─█─`:`
#>   ├─1
#>   └─10
lobstr::ast(mean(!!(1:10)))
#> █─mean
#> └─<inline integer>
```

In the expression `mean(!!(1:10))` the call `1:10` is evaluated to an integer vector, while still being a call object in `mean(1:10)`.

The first version (`mean(1:10)`) seems more natural. It captures lazy evaluation, with a promise that is evaluated when the function is called. The second version (`mean(!!(1:10))`) inlines a vector directly into a call.

19.6 ... (dot-dot-dot)

Q1: One way to implement `exec()` is shown below. Describe how it works. What are the key ideas?

```
exec <- function(f, ..., .env = caller_env()) {
  args <- list2(...)
  do.call(f, args, envir = .env)
}
```

A: `exec()` takes a function (`f`), its arguments (`...`) and an environment (`.env`) as input. This allows to construct a call from `f` and `...` and evaluate this call in the supplied environment. As the `...` argument is handled via `list2()`, `exec()` supports tidy dots (quasiquotation), which means that arguments and names (on the left-hand side of `:=`) can be unquoted via `!!` and `!!!`.

Q2: Carefully read the source code for `interaction()`, `expand.grid()`, and `par()`. Compare and contrast the techniques they use for switching between dots and list behaviour.

A: All three functions capture the dots via `args <- list(...)`.

`interaction()` computes factor interactions between the captured input factors by iterating over the `args`. When a list is provided this is detected via `length(args) == 1 && is.list(args[[1]])` and one level of the list is stripped through `args <- args[[1]]`. The rest of the function's code doesn't differentiate further between list and dots behaviour.

```
# Both calls create the same output
interaction(     a = c("a", "b", "c", "d"), b = c("e", "f"))   # dots
#> [1] a.e b.f c.e d.f
#> Levels: a.e b.e c.e d.e a.f b.f c.f d.f
interaction(list(a = c("a", "b", "c", "d"), b = c("e", "f")))  # list
#> [1] a.e b.f c.e d.f
#> Levels: a.e b.e c.e d.e a.f b.f c.f d.f
```

`expand.grid()` uses the same strategy and also assigns `args <- args[[1]]` in case of `length(args) == 1 && is.list(args[[1]])`.

`par()` does the most pre-processing to ensure a valid structure of the `args` argument. When no dots are provided (`!length(args)`) it creates a list of arguments from an internal character vector (partly depending on its `no.readonly` argument). Further, given that all elements of `args` are character vectors (`all(unlist(lapply(args, is.character)))`), `args` is turned into a list via `as.list(unlist(args))` (this flattens nested lists). Similar to the other functions, one level of `args` gets stripped via `args <- args[[1L]]`, when `args` is of length one and its first element is a list.

Q3: Explain the problem with this definition of `set_attr()`

```
set_attr <- function(x, ...) {
  attr <- rlang::list2(...)
  attributes(x) <- attr
  x
}
set_attr(1:10, x = 10)
#> Error in attributes(x) <- attr: attributes must be named
```

A: `set_attr()` expects an object named `x` and its attributes, supplied via the dots. Unfortunately, this prohibits us to provide attributes named `x` as these would collide with the argument name of our object. Even omitting the object's argument name doesn't help in this case — as can be seen in the example where the object is consequently treated as an unnamed attribute.

However, we may name the first argument `.x`, which seems clearer and less likely to invoke errors. In this case `1:10` will get the (named) attribute `x = 10` assigned:

```
set_attr <- function(.x, ...) {
  attr <- rlang::list2(...)

  attributes(.x) <- attr
  .x
}

set_attr(1:10, x = 10)
#>  [1]  1  2  3  4  5  6  7  8  9 10
#> attr(,"x")
#> [1] 10
```

19.7 Case studies

Q1: In the linear-model example, we could replace the `expr()` in `reduce(summands, ~ expr(!!.x + !!.y))` with `call2()`: `reduce(summands, call2, "+")`. Compare and contrast the two approaches. Which do you think is easier to read?

A: We would consider the first version to be more readable. There seems to be a little more boilerplate code at first, but the unquoting syntax is very readable. Overall, the whole expression seems more explicit and less complex.

Q2: Re-implement the Box-Cox transform defined below using unquoting and `new_function()`:

```
bc <- function(lambda) {
  if (lambda == 0) {
    function(x) log(x)
  } else {
    function(x) (x ^ lambda - 1) / lambda
  }
}
```

A: Here new_function() allows us to create a function factory using tidy evaluation.

```
bc2 <- function(lambda) {
  lambda <- enexpr(lambda)

  if (!!lambda == 0) {
    new_function(exprs(x = ), expr(log(x)))
  } else {
    new_function(exprs(x = ), expr((x ^ (!!lambda) - 1) / !!lambda))
  }
}

bc2(0)
#> function (x)
#> log(x)
#> <environment: 0x562ef541ac40>
bc2(2)
#> function (x)
#> (x^2 - 1)/2
#> <environment: 0x562ef54720b8>
bc2(2)(2)
#> [1] 1.5
```

Q3: Re-implement the simple compose() defined below using quasiquotation and new_function():

```
compose <- function(f, g) {
  function(...) f(g(...))
}
```

A: The implementation is fairly straightforward, even though a lot of parentheses are required:

```
compose2 <- function(f, g) {
  f <- enexpr(f)
  g <- enexpr(g)

  new_function(exprs(... = ), expr((!!f)((!!g)(...))))
}

compose(sin, cos)
#> function(...) f(g(...))
```

```
#> <environment: 0x562ef4207d78>
compose(sin, cos)(pi)
#> [1] -0.841

compose2(sin, cos)
#> function (...)
#> sin(cos(...))
#> <environment: 0x562ef4403098>
compose2(sin, cos)(pi)
#> [1] -0.841
```

20

Evaluation

Prerequisites

On our journey through R's metaprogramming, we continue to use the functions from the {rlang} package.

```
library(rlang)
```

20.2 Evaluation basics

Q1: Carefully read the documentation for `source()`. What environment does it use by default? What if you supply `local = TRUE`? How do you provide a custom environment?

A: By default, `source()` uses the global environment (`local = FALSE`). A specific evaluation environment may be chosen, by passing it explicitly to the `local` argument. To use current environment (i.e. the calling environment of `source()`) set `local = TRUE`.

```
# Create a temporary, sourcable R script that prints x
tmp_file <- tempfile()
writeLines("print(x)", tmp_file)

# Set `x` globally
x <- "global environment"
env2 <- env(x = "specified environment")

locate_evaluation <- function(file, local) {
  x <- "local environment"
```

DOI: 10.1201/9781003175414-20

```r
  source(file, local = local)
}

# Where will source() evaluate the code?
locate_evaluation(tmp_file, local = FALSE)  # default
#> [1] "global environment"
locate_evaluation(tmp_file, local = env2)
#> [1] "specified environment"
locate_evaluation(tmp_file, local = TRUE)
#> [1] "local environment"
```

Q2: Predict the results of the following lines of code:

```r
eval(expr(eval(expr(eval(expr(2 + 2))))))        # (1)
eval(eval(expr(eval(expr(eval(expr(2 + 2))))))) # (2)
expr(eval(expr(eval(expr(eval(expr(2 + 2))))))) # (3)
```

A: Let's look at a quote from the first edition of *Advanced R* (`http://adv-r.had.co.nz/Computing-on-the-language.html#subset`):

> "`expr()` and `eval()` are opposites. [...] each `eval()` peels off one layer of `expr()`'s".

In general, `eval(expr(x))` evaluates to `x`. Therefore, (1) evaluates to $2+2=4$. Adding another `eval()` doesn't have impact here. So, also (2) evaluates to `4`. However, when wrapping (1) into `expr()` the whole expression will be quoted.

```r
eval(expr(eval(expr(eval(expr(2 + 2))))))        # (1)
#> [1] 4
eval(eval(expr(eval(expr(eval(expr(2 + 2))))))) # (2)
#> [1] 4
expr(eval(expr(eval(expr(eval(expr(2 + 2))))))) # (3)
#> eval(expr(eval(expr(eval(expr(2 + 2))))))
```

Q3: Fill in the function bodies below to re-implement `get()` using `sym()` and `eval()`, and `assign()` using `sym()`, `expr()`, and `eval()`. Don't worry about the multiple ways of choosing an environment that `get()` and `assign()` support; assume that the user supplies it explicitly.

```r
# name is a string
get2 <- function(name, env) {}
assign2 <- function(name, value, env) {}
```

A: We reimplement these two functions using tidy evaluation. We turn the string name into a symbol, then evaluate it:

```
get2 <- function(name, env = caller_env()) {
  name_sym <- sym(name)
  eval(name_sym, env)
}

x <- 1
get2("x")
#> [1] 1
```

To build the correct expression for the value assignment, we unquote using `!!`.

```
assign2 <- function(name, value, env = caller_env()) {
  name_sym <- sym(name)
  assign_expr <- expr(!!name_sym <- !!value)
  eval(assign_expr, env)
}

assign2("x", 4)
x
#> [1] 4
```

Q4: Modify `source2()` so it returns the result of *every* expression, not just the last one. Can you eliminate the for loop?

A: The code for `source2()` was given in *Advanced R* as:

```
source2 <- function(path, env = caller_env()) {
  file <- paste(readLines(path, warn = FALSE), collapse = "\n")
  exprs <- parse_exprs(file)

  res <- NULL
  for (i in seq_along(exprs)) {
    res <- eval(exprs[[i]], env)
  }

  invisible(res)
}
```

In order to highlight the modifications in our new `source2()` function, we've preserved the differing code from the former `source2()` in a comment.

```
source2 <- function(path, env = caller_env()) {
  file <- paste(readLines(path, warn = FALSE), collapse = "\n")
  exprs <- parse_exprs(file)

  # res <- NULL
  # for (i in seq_along(exprs)) {
  #   res[[i]] <- eval(exprs[[i]], env)
  # }

  res <- purrr::map(exprs, eval, env)

  invisible(res)
}
```

Let's create a file and test `source2()`. Keep in mind that `<-` returns invisibly.

```
tmp_file <- tempfile()
writeLines(
  "x <- 1
       x
       y <- 2
       y  # some comment",
  tmp_file
)

(source2(tmp_file))
#> [[1]]
#> [1] 1
#>
#> [[2]]
#> [1] 1
#>
#> [[3]]
#> [1] 2
#>
#> [[4]]
#> [1] 2
```

Q5: We can make `base::local()` slightly easier to understand by spreading it over multiple lines:

```
local3 <- function(expr, envir = new.env()) {
  call <- substitute(eval(quote(expr), envir))
```

```
  eval(call, envir = parent.frame())
}
```

Explain how `local()` works in words. (Hint: you might want to `print(call)` to help understand what `substitute()` is doing, and read the documentation to remind yourself what environment `new.env()` will inherit from.)

A: Let's follow the advice and add `print(call)` inside of `local3()`:

```
local3 <- function(expr, envir = new.env()) {
  call <- substitute(eval(quote(expr), envir))
  print(call)
  eval(call, envir = parent.frame())
}
```

The first line generates a call to `eval()`, because `substitute()` operates in the current evaluation argument. However, this doesn't matter here, as both, `expr` and `envir` are promises and therefore "the expression slots of the promises replace the symbols", from `?substitute`.

```
local3({
  x <- 10
  x * 2
})
#> eval(quote({
#>     x <- 10
#>     x * 2
#> }), new.env())
#> [1] 20
```

Next, `call` will be evaluated in the caller environment (aka the parent frame). Given that `call` contains another call `eval()` why does this matter? The answer is subtle: this outer environment determines where the bindings for `eval`, `quote`, and `new.env` are found.

```
eval(quote({
  x <- 10
  x * 2
}), new.env())
#> [1] 20
exists("x")
#> [1] TRUE
```

20.3 Quosures

Q1: Predict what evaluating each of the following quosures will return if evaluated.

```
q1 <- new_quosure(expr(x), env(x = 1))
q1
#> <quosure>
#> expr: ^x
#> env:  0x564a754b22a8

q2 <- new_quosure(expr(x + !!q1), env(x = 10))
q2
#> <quosure>
#> expr: ^x + (^x)
#> env:  0x564a75cee228

q3 <- new_quosure(expr(x + !!q2), env(x = 100))
q3
#> <quosure>
#> expr: ^x + (^x + (^x))
#> env:  0x564a75eac2f0
```

A: Each quosure is evaluated in its own environment, so `x` is bound to a different value for each time. This leads us to:

```
eval_tidy(q1)
#> [1] 1
eval_tidy(q2)
#> [1] 11
eval_tidy(q3)
#> [1] 111
```

Q2: Write an `enenv()` function that captures the environment associated with an argument. (Hint: this should only require two function calls.)

A: A quosure captures both the expression and the environment. From a quosure, we can access the environment with the help of `get_env()`.

```
enenv <- function(x) {
  get_env(enquo(x))
}
```

```
# Test
enenv(x)
#> <environment: R_GlobalEnv>

# Test if it also works within functions
capture_env <- function(x) {
  enenv(x)
}
capture_env(x)
#> <environment: 0x564a767bc318>
```

20.4 Data masks

Q1: Why did I use a for loop in `transform2()` instead of `map()`? Consider `transform2(df, x = x * 2, x = x * 2)`.

A: `transform2()` was defined in *Advanced R* as:

```
transform2 <- function(.data, ...) {
  dots <- enquos(...)

  for (i in seq_along(dots)) {
    name <- names(dots)[[i]]
    dot <- dots[[i]]

    .data[[name]] <- eval_tidy(dot, .data)
  }

  .data
}
```

A for loop applies the processing steps regarding `.data` iteratively. This includes updating `.data` and reusing the same variable names. This makes it possible to apply transformations sequentially, so that subsequent transformations can refer to columns that were just created.

Q2: Here's an alternative implementation of `subset2()`:

```
subset3 <- function(data, rows) {
  rows <- enquo(rows)
  eval_tidy(expr(data[!!rows, , drop = FALSE]), data = data)
}

df <- data.frame(x = 1:3)
subset3(df, x == 1)
```

Compare and contrast `subset3()` to `subset2()`. What are its advantages and disadvantages?

A: Let's take a closer look at `subset2()` first:

```
subset2 <- function(data, rows) {
  rows <- enquo(rows)
  rows_val <- eval_tidy(rows, data)
  stopifnot(is.logical(rows_val))

  data[rows_val, , drop = FALSE]
}
```

`subset2()` provides an additional logical check, which is missing from `subset3()`. Here, `rows` is evaluated in the context of `data`, which results in a logical vector. Afterwards only `[` needs to be used for subsetting.

```
# subset2() evaluation
(rows_val <- eval_tidy(quo(x == 1), df))
#> [1]  TRUE FALSE FALSE
df[rows_val, , drop = FALSE]
#>   x
#> 1 1
```

With `subset3()` both of these steps occur in a single line (which is probably closer to what one would produce by hand). This means that the subsetting is also evaluated in the context of the data mask.

```
# subset3() evaluation
eval_tidy(expr(df[x == 1, , drop = FALSE]), df)
#>   x
#> 1 1
```

This is shorter (but probably also less readable) because the evaluation and the subsetting take place in the same expression. However, it may introduce unwanted errors, if the data mask contains an element named "data", as the objects from the data mask take precedence over arguments of the function.

```
df <- data.frame(x = 1:3, data = 1)
subset2(df, x == 1)
#>   x data
#> 1 1    1
subset3(df, x == 1)
#> Error in data[~x == 1, , drop = FALSE]: incorrect number of
#> dimensions
```

Q3: The following function implements the basics of `dplyr::arrange()`. Annotate each line with a comment explaining what it does. Can you explain why `!!.na.last` is strictly correct, but omitting the `!!` is unlikely to cause problems?

```
arrange2 <- function(.df, ..., .na.last = TRUE) {
  args <- enquos(...)

  order_call <- expr(order(!!!args, na.last = !!.na.last))

  ord <- eval_tidy(order_call, .df)
  stopifnot(length(ord) == nrow(.df))

  .df[ord, , drop = FALSE]
}
```

A: `arrange2()` basically reorders a data frame by one or more of its variables. As `arrange2()` allows to provide the variables as expressions (via `...`), these need to be quoted first. Afterwards they are used to build up an `order()` call, which is then evaluated in the context of the data frame. Finally, the data frame is reordered via integer subsetting. Let's take a closer look at the source code:

```
arrange2 <- function(.df, ..., .na.last = TRUE) {
  # Capture and quote arguments, which determine the order
  args <- enquos(...)

  # `!!!`: unquote-splice arguments into order()
  # `!!.na.last`: pass option for treatment of NAs to order()
  # return expression-object
  order_call <- expr(order(!!!args, na.last = !!.na.last))

  # Evaluate order_call within .df
```

```
  ord <- eval_tidy(order_call, .df)
  # Ensure that no rows are dropped
  stopifnot(length(ord) == nrow(.df))

  # Reorder rows via integer subsetting
  .df[ord, , drop = FALSE]
}
```

By using `!!.na.last` the `.na.last` argument is unquoted when the `order()` call is built. This way, the `na.last` argument is already correctly specified (typically `TRUE`, `FALSE` or `NA`).

Without the unquoting, the expression would read `na.last = .na.last` and the value for `.na.last` would still need to be looked up and found. Because these computations take place inside of the function's execution environment (which contains `.na.last`), this is unlikely to cause problems.

```
# The effect of unquoting .na.last
.na.last <- FALSE
expr(order(..., na.last = !!.na.last))
#> order(..., na.last = FALSE)
expr(order(..., na.last = .na.last))
#> order(..., na.last = .na.last)
```

20.5 Using tidy evaluation

Q1: I've included an alternative implementation of `threshold_var()` below. What makes it different to the approach I used above? What makes it harder?

```
threshold_var2 <- function(df, var, val) {
  var <- ensym(var)

  subset2(df, `$`(.data, !!var) >= !!val)
}
```

A: Let's compare this approach to the original implementation:

```
threshold_var <- function(df, var, val) {
  var <- as_string(ensym(var))
  subset2(df, .data[[var]] >= !!val)
}
```

We can see that `threshold_var2()` no longer coerces the symbol to a string. Therefore `$` instead of `[[` can be used for subsetting. Initially we suspected partial matching would be introduced by `$`, but `.data` deliberately avoids this problem.

The prefix call to `$()` is less common than infix-subsetting using `[[`, but ultimately both functions behave the same.

```
df <- data.frame(x = 1:10)
threshold_var(df, x, 8)
#>     x
#> 8   8
#> 9   9
#> 10 10
threshold_var2(df, x, 8)
#>     x
#> 8   8
#> 9   9
#> 10 10
```

20.6 Base evaluation

Q1: Why does this function fail?

```
lm3a <- function(formula, data) {
  formula <- enexpr(formula)

  lm_call <- expr(lm(!!formula, data = data))
  eval(lm_call, caller_env())
}
lm3a(mpg ~ disp, mtcars)$call
#> Error in model.frame.default(formula = mpg ~ disp, data = data,
#> drop.unused.levels = TRUE): 'data' must be a data.frame,
#> environment, or list
```

A: In this function, `lm_call` is evaluated in the caller environment, which happens to be the global environment. In this environment, the name `data` is bound to `utils::data`. To fix the error, we can either set the evaluation environment to the function's execution environment or unquote the `data` argument when building the call to `lm()`.

```
# Change evaluation environment
lm3b <- function(formula, data) {
  formula <- enexpr(formula)

  lm_call <- expr(lm(!!formula, data = data))
  eval(lm_call, current_env())
}

lm3b(mpg ~ disp, mtcars)$call
#> lm(formula = mpg ~ disp, data = data)
lm3b(mpg ~ disp, data)$call  #reproduces original error
#> Error in model.frame.default(formula = mpg ~ disp, data = data,
#> drop.unused.levels = TRUE): 'data' must be a data.frame,
#> environment, or list
```

When we want to unquote an argument within a function, we first need to capture the user-input (by `enexpr()`).

```
# Unquoting data-argument
lm3c <- function(formula, data) {
  formula <- enexpr(formula)
  data_quo <- enexpr(data)

  lm_call <- expr(lm(!!formula, data = !!data_quo))
  eval(lm_call, caller_env())
}
lm3c(mpg ~ disp, mtcars)$call
#> lm(formula = mpg ~ disp, data = mtcars)
```

Q2: When model building, typically the response and data are relatively constant while you rapidly experiment with different predictors. Write a small wrapper that allows you to reduce duplication in the code below.

```
lm(mpg ~ disp, data = mtcars)
lm(mpg ~ I(1 / disp), data = mtcars)
lm(mpg ~ disp * cyl, data = mtcars)
```

A: In our wrapper `lm_wrap()`, we provide `mpg` and `mtcars` as default response and data. This seems to give us a good mix of usability and flexibility.

```
lm_wrap <- function(pred, resp = mpg, data = mtcars,
                    env = caller_env()) {
  pred <- enexpr(pred)
  resp <- enexpr(resp)
  data <- enexpr(data)

  formula <- expr(!!resp ~ !!pred)
  lm_call <- expr(lm(!!formula, data = !!data))
  eval(lm_call, envir = env)
}

# Test if the output looks ok
lm_wrap(I(1 / disp) + disp * cyl)
#>
#> Call:
#> lm(formula = mpg ~ I(1/disp) + disp * cyl, data = mtcars)
#>
#> Coefficients:
#> (Intercept)    I(1/disp)         disp          cyl     disp:cyl
#>   -1.22e+00     1.85e+03     7.68e-02     1.18e+00    -9.14e-03

# Test if the result is identical to calling lm() directly
identical(
  lm_wrap(I(1 / disp) + disp * cyl),
  lm(mpg ~ I(1 / disp) + disp * cyl, data = mtcars)
)
#> [1] TRUE
```

Q3: Another way to write `resample_lm()` would be to include the resample expression `(data[sample(nrow(data), replace = TRUE), , drop = FALSE])` in the data argument. Implement that approach. What are the advantages? What are the disadvantages?

A: Different versions of `resample_lm()` were given in *Advanced R*. However, none of them implemented the resampling within the function argument.

Different versions of `resample_lm()` (`resample_lm0()`, `resample_lm1()`, `resample_lm2()`) were specified in *Advanced R*. However, in none of these versions was the resampling step implemented in any of the arguments.

This approach takes advantage of R's lazy evaluation of function arguments, by moving the resampling step into the argument definition. The user passes the data to the function, but only a permutation of this data (`resample_data`) will be used.

```
resample_lm <- function(
  formula, data,
  resample_data = data[sample(nrow(data), replace = TRUE), ,
                       drop = FALSE],
  env = current_env()) {

  formula <- enexpr(formula)

  lm_call <- expr(lm(!!formula, data = resample_data))
  expr_print(lm_call)
  eval(lm_call, env)
}

df <- data.frame(x = 1:10, y = 5 + 3 * (1:10) + round(rnorm(10), 2))
(lm_1 <- resample_lm(y ~ x, data = df))
#> lm(y ~ x, data = resample_data)
#>
#> Call:
#> lm(formula = y ~ x, data = resample_data)
#>
#> Coefficients:
#> (Intercept)            x
#>        4.85         3.02
lm_1$call
#> lm(formula = y ~ x, data = resample_data)
```

With this approach the evaluation needs to take place within the function's environment, because the resampled dataset (defined as a default argument) will only be available in the function environment.

Overall, putting an essential part of the pre-processing outside of the functions body is not common practice in R. Compared to the unquoting-implementation (`resample_lm1()` in *Advanced R*), this approach captures the model-call in a more meaningful way. This approach will also lead to a new resample every time you `update()` the model.

21

Translating R code

Prerequisites

In this chapter we combine R's metaprogramming and functional programming capabilities and therefore load both the {rlang} and the {purrr} package.

```
library(rlang)
library(purrr)
```

21.2 HTML

Q1: The escaping rules for <script> tags are different because they contain JavaScript, not HTML. Instead of escaping angle brackets or ampersands, you need to escape </script> so that the tag isn't closed too early. For example, `script("'</script>'")`, shouldn't generate this:

```
<script>'</script>'</script>
```

But

```
<script>'<\/script>'</script>
```

Adapt the `escape()` to follow these rules when a new argument `script` is set to `TRUE`.

A: We are asked to implement a special case of escaping for the <script> tag. At first we will revisit the relevant functions provided in *Advanced R* and confirm that our code reliably escapes for tags like <p> and <b> but doesn't

DOI: 10.1201/9781003175414-21

escape correctly for the `<script>` tag. Then we modify the `escape()` and `tag()` functions to redefine the `<script>` tag and confirm that all defined tags now escape correctly.

Note that the `<style>` tag, which contains styling information in CSS, follows the same escaping rules as the `<script>` tag. We therefore implement the desired escaping for the `<style>` tag function also.

Let's start by loading the relevant code from *Advanced R* first.

```
# Escaping
html <- function(x) structure(x, class = "advr_html")

print.advr_html <- function(x, ...) {
  out <- paste0("<HTML> ", x)
  cat(paste(strwrap(out), collapse = "\n"), "\n", sep = "")
}

escape <- function(x) UseMethod("escape")

escape.character <- function(x) {
  x <- gsub("&", "&amp;", x)
  x <- gsub("<", "&lt;", x)
  x <- gsub(">", "&gt;", x)

  html(x)
}

escape.advr_html <- function(x) x

# Basic tag functions
dots_partition <- function(...) {
  dots <- list2(...)

  if (is.null(names(dots))) {
    is_named <- rep(FALSE, length(dots))
  } else {
    is_named <- names(dots) != ""
  }

  list(
    named = dots[is_named],
    unnamed = dots[!is_named]
  )
}
```

```
# html_attributes() function from the GitHub repository of Advanced R
# https://github.com/hadley/adv-r/blob/master/dsl-html-attributes.r

html_attributes <- function(list) {
  if (length(list) == 0) return("")

  attr <- map2_chr(names(list), list, html_attribute)
  paste0(" ", unlist(attr), collapse = "")
}

html_attribute <- function(name, value = NULL) {
  if (length(value) == 0) return(name) # for attributes with no value
  if (length(value) != 1) stop("`value` must be NULL or length 1")
  if (is.logical(value)) {
    # Convert T and F to true and false
    value <- tolower(value)
  } else {
    value <- escape_attr(value)
  }
  paste0(name, "='", value, "'")
}

escape_attr <- function(x) {
  x <- escape.character(x)
  x <- gsub("\'", '&#39;', x)
  x <- gsub("\"", '&quot;', x)
  x <- gsub("\r", '&#13;', x)
  x <- gsub("\n", '&#10;', x)
  x
}

# Tag functions
tag <- function(tag) {
  new_function(
    exprs(... = ),
    expr({
      dots <- dots_partition(...)
      attribs <- html_attributes(dots$named)
      children <- map_chr(dots$unnamed, escape)

      html(paste0(
        !!paste0("<", tag), attribs, ">",
        paste(children, collapse = ""),
```

```
        !!paste0("</", tag, ">")
      ))
    }),
    caller_env()
  )
}
```

This code escapes the <p> and <b> tags correctly, but doesn't achieve the desired behaviour for the <script> tag yet:

```
p <- tag("p")
b <- tag("b")

identical(
  p("&","and <", b("& > will be escaped")) %>%
    as.character(),
  "<p>&amp;and &lt;<b>&amp; &gt; will be escaped</b></p>"
)
#> [1] TRUE

script <- tag("script")

identical(
  script("Don't escape &, <, > - escape </script> and </style>") %>%
    as.character(),
  paste("<script>Don't escape &, <, >",
        "- escape <\\/script> and <\\/style></script>")
)
#> [1] FALSE
```

We implement the desired change and add the optional argument `script` to the `escape()` and the `tag()` functions (default: `script = FALSE`). The argument has to be added for all methods of the `escape()` generic.

```
escape <- function(x, script = FALSE) UseMethod("escape")

escape.character <- function(x, script = FALSE) {

  if (script) {
    x <- gsub("</script>", "<\\/script>", x, fixed = TRUE)
    x <- gsub("</style>",  "<\\/style>",  x, fixed = TRUE)
  } else {
    x <- gsub("&", "&amp;", x)
```

```
    x <- gsub("<", "&lt;", x)
    x <- gsub(">", "&gt;", x)
  }

  html(x)
}

escape.advr_html <- function(x, script = FALSE) x


tag <- function(tag, script = FALSE) {

  new_function(
    exprs(... = ),
    expr({
      dots <- dots_partition(...)
      attribs <- html_attributes(dots$named)
      children <- map_chr(dots$unnamed, escape, script = !!script)
      html(paste0(
        !!paste0("<", tag), attribs, ">",
        paste(children, collapse = ""),
        !!paste0("</", tag, ">")
      ))
    }),
    caller_env()
  )
}
```

Finally, we create new `<p>`, `<b>` and `<script>` tag functions, which now pass their escaping tests.

```
p <- tag("p")
b <- tag("b")

identical(
  p("&","and <", b("& > will be escaped")) %>%
    as.character(),
  "<p>&amp;and &lt;<b>&amp; &gt; will be escaped</b></p>"
)
#> [1] TRUE

script <- tag("script", script = TRUE)
style  <- tag("style" , script = TRUE)
```

```
identical(
  script("Don't escape &, <, > - escape </script> and </style>") %>%
    as.character(),
  paste("<script>Don't escape &, <, >",
        "- escape <\\/script> and <\\/style></script>")
)
#> [1] TRUE

script("Don't escape &, <, > - escape </script> and </style>")
#> <HTML> <script>Don't escape &, <, > - escape <\/script> and
#> <\/style></script>
```

Q2: The use of `...` for all functions has some big downsides. There's no input validation and there will be little information in the documentation or autocomplete about how they are used in the function. Create a new function that, when given a named list of tags and their attribute names (like below), creates tag functions with named arguments.

```
list(
  a = c("href"),
  img = c("src", "width", "height")
)
```

All tags should get `class` and `id` attributes.

A: This exercise requires a function factory: The named list of attribute names will be extended (by `class` and `id`) and mapped to function arguments. These will default to `NULL`, so that the user isn't forced to provide them.

When creating the tag functions itself we use `check_dots_unnamed()` from the `{ellipsis}` package to ensure named arguments correspond to the expected values (and are not created by some spelling mistake). After that we follow the logic from the `tag()` function factory above.

To keep the focus on the key ideas, we ignore special cases like `<script>`, `<style>` and void tags in this solution (even if this leads to an incorrect tag function for the `<img>` tag).

```
tag_factory <- function(tag, tag_attrs) {
  attrs <- c("class", "id", tag_attrs)

  attr_args <- set_names(rep(list(NULL), length(attrs)), attrs)
  attr_list <- call2("list", !!!syms(set_names(attrs)))
```

```
  new_function(
    exprs(... = , !!!attr_args),
    expr({
      ellipsis::check_dots_unnamed()

      attribs <- html_attributes(compact(!!attr_list))
      dots <- compact(list(...))
      children <- map_chr(dots, escape)

      html(paste0(
        !!paste0("<", tag), attribs, ">",
        paste(children, collapse = ""),
        !!paste0("</", tag, ">")
      ))
    })
  )
}
```

To validate our new function factory, we modify the `with_html()` example from *Advanced R* to work with our newly created `a()` and `img()` tag functions.

```
tag_list <- list(
  a = c("href"),
  img = c("src", "width", "height")
)

tags <- map2(names(tag_list), unname(tag_list), tag_factory) %>%
  set_names(names(tag_list))

with_tags <- function(code) {
  code <- enquo(code)
  eval_tidy(code, tags)
}

with_tags(
  a(
    img("Correct me if I am wrong", id = "second"),
    href = "https://github.com/Tazinho/Advanced-R-Solutions/issues",
    id = "first"
  )
)
#> <HTML> <a id='first'
```

```
#> href='https://github.com/Tazinho/Advanced-R-Solutions/issues'><img
#> id='second'>Correct me if I am wrong</img></a>
```

Q3: Reason about the following code that calls `with_html()` referencing objects from the environment. Will it work or fail? Why? Run the code to verify your predictions.

```
greeting <- "Hello!"
with_html(p(greeting))

p <- function() "p"
address <- "123 anywhere street"
with_html(p(address))
```

A: First, we rerun the relevant code from *Advanced R* to define `with_html()`. Note that we skip the code for void tags, as none of them appear in the code chunk from this exercise.

```
tags <- c(
  "a", "abbr", "address", "article", "aside", "audio",
  "b", "bdi", "bdo", "blockquote", "body", "button", "canvas",
  "caption", "cite", "code", "colgroup", "data", "datalist",
  "dd", "del", "details", "dfn", "div", "dl", "dt", "em",
  "eventsource", "fieldset", "figcaption", "figure", "footer",
  "form", "h1", "h2", "h3", "h4", "h5", "h6", "head", "header",
  "hgroup", "html", "i", "iframe", "ins", "kbd", "label",
  "legend", "li", "mark", "map", "menu", "meter", "nav",
  "noscript", "object", "ol", "optgroup", "option", "output",
  "p", "pre", "progress", "q", "ruby", "rp", "rt", "s", "samp",
  "script", "section", "select", "small", "span", "strong",
  "style", "sub", "summary", "sup", "table", "tbody", "td",
  "textarea", "tfoot", "th", "thead", "time", "title", "tr",
  "u", "ul", "var", "video"
)

html_tags <- tags %>% set_names() %>% map(tag)

with_html <- function(code) {
  code <- enquo(code)
  eval_tidy(code, html_tags)
}
```

Now, let us briefly repeat, that `with_html()` was introduced to evaluate tag functions from within a list. Otherwise, defining some tag functions like `body()`, `source()`, `summary()` etc. within the global environment would collide with base R functions with the same name. To prevent this the DSL code wrapped in `with_html()` is evaluated within the "context" of `html_tags`, which was provided as a data mask to `eval_tidy()`. As `?rlang::as_data_mask` mentions: "Objects in the mask have precedence over objects in the environment".

Therefore, `p()` refers to the tag function from `html_tags` within both examples from the exercise. However, as `address` is not only a string within the global environment, but also a tag function within `html_tags` (the `<address>` HTML tag may be used to provide contact information on an HTML page), `p()` operates on `address()` in the second example. This correctly leads to an error as we haven't implemented an `escape.function()` method.

```
greeting <- "Hello!"
with_html(p(greeting))
#> <HTML> <p>Hello!</p>

p <- function() "p"
address <- "123 anywhere street"
with_html(p(address))
#> Error in UseMethod("escape"): no applicable method for 'escape'
#> applied to an object of class "function"
```

Q4: Currently the HTML doesn't look terribly pretty, and it's hard to see the structure. How could you adapt `tag()` to do indenting and formatting? (You may need to do some research into block and inline tags.)

A: First, let us load all relevant functions from *Advanced R*:

```
tag <- function(tag) {
  new_function(
    exprs(... = ),
    expr({
      dots <- dots_partition(...)
      attribs <- html_attributes(dots$named)
      children <- map_chr(dots$unnamed, escape)
      html(paste0(
        !!paste0("<", tag), attribs, ">",
        paste(children, collapse = ""),
        !!paste0("</", tag, ">")
      ))
    }),
```

```
    caller_env()
  )
}

void_tag <- function(tag) {
  new_function(
    exprs(... = ),
    expr({
      dots <- dots_partition(...)
      if (length(dots$unnamed) > 0) {
        stop(
          !!paste0("<", tag, "> must not have unnamed arguments"),
          call. = FALSE
        )
      }

      attribs <- html_attributes(dots$named)

      html(paste0(!!paste0("<", tag), attribs, " />"))
    }),
    caller_env()
  )
}

tags <- c(
  "a", "abbr", "address", "article", "aside", "audio", "b",
  "bdi", "bdo", "blockquote", "body", "button", "canvas",
  "caption", "cite", "code", "colgroup", "data", "datalist",
  "dd", "del", "details", "dfn", "div", "dl", "dt", "em",
  "eventsource", "fieldset", "figcaption", "figure", "footer",
  "form", "h1", "h2", "h3", "h4", "h5", "h6", "head", "header",
  "hgroup", "html", "i", "iframe", "ins", "kbd", "label", "legend",
  "li", "mark", "map", "menu", "meter", "nav", "noscript", "object",
  "ol", "optgroup", "option", "output", "p", "pre", "progress", "q",
  "ruby", "rp", "rt", "s", "samp", "script", "section", "select",
  "small", "span", "strong", "style", "sub", "summary", "sup",
  "table", "tbody", "td", "textarea", "tfoot", "th", "thead",
  "time", "title", "tr", "u", "ul", "var", "video"
)

void_tags <- c(
  "area", "base", "br", "col", "command", "embed", "hr", "img",
  "input", "keygen", "link", "meta", "param", "source",
  "track", "wbr"
```

```
)

html_tags <- c(
  tags %>% set_names() %>% map(tag),
  void_tags %>% set_names() %>% map(void_tag)
)

with_html <- function(code) {
  code <- enquo(code)
  eval_tidy(code, html_tags)
}
```

Now, let's look at the example from above:

```
with_html(
  body(
    h1("A heading", id = "first"),
    p("Some text &", b("some bold text.")),
    img(src = "myimg.png", width = 100, height = 100)
  )
)
#> <HTML> <body><h1 id='first'>A heading</h1><p>Some text
#> &amp;<b>some bold text.</b></p><img src='myimg.png'
#> width='100' height='100' /></body>
```

The formatting consists of only one long line of code. This output makes it difficult to check the content of the HTML code and its correctness.

What kind of formatting would we prefer instead? Google's HTML style guide (`https://google.github.io/styleguide/htmlcssguide.html#HTML_Formatting_Rules`) suggests *indentation* by 2 spaces and *new lines* for every block, list, or table element. There are other recommendations, but we will keep things simple and will be satisfied with the following output.

```
<body>
  <h1 id='first'>A heading</h1>
  <p>Some text &amp;<b>some bold text.</b></p>
  <img src='myimg.png'width='100' height='100' />
</body>
```

First we adjust the `print.advr_html()` method, removing `strwrap()` function, because this will re-wrap the HTML, making it harder to understand what's happening.

```
html <- function(x) structure(x, class = "advr_html")

print.advr_html <- function(x, ...) {
  cat(paste("<HTML>", x, sep = "\n"))
}
```

In our desired output we can see that the content of the body-function requires different formatting than the other tag-functions. We will therefore create a new format_code() function, that allows for optional indentation and line breaks.

```
indent <- function(x) {
  paste0("  ", gsub("\n", "\n  ", x))
}

format_code <- function(children, indent = FALSE) {
  if (indent) {
    paste0("\n", paste0(indent(children), collapse = "\n"), "\n")
  } else {
    paste(children, collapse = "")
  }
}
```

We adjust the body function to include the format_code() helper. (This could also be approached programmatically in the tag function factory.)

```
html_tags$body <- function(...) {
  dots <- dots_partition(...)
  attribs <- html_attributes(dots$named)
  children <- map_chr(dots$unnamed, escape)

  html(paste0(
    "<body", attribs, ">",
    format_code(children, indent = TRUE),
    "</body>"
  ))
}
```

The resulting output is much more satisfying.

```
with_html(
  body(
```

```
    h1("A heading", id = "first"),
    p("Some text &", b("some bold text.")),
    img(src = "myimg.png", width = 100, height = 100)
  )
)
#> <HTML>
#> <body>
#>   <h1 id='first'>A heading</h1>
#>   <p>Some text &amp;<b>some bold text.</b></p>
#>   <img src='myimg.png' width='100' height='100' />
#> </body>
```

21.3 LaTeX

Q1: Add escaping. The special symbols that should be escaped by adding a backslash in front of them are `\`, `$`, and `%`. Just as with HTML, you'll need to make sure you don't end up double-escaping. So, you'll need to create a small S3 class and then use that in function operators. That will also allow you to embed arbitrary LaTeX if needed.

A: Currently our `to_math()` function generates the following output:

```
to_math(`$`)
#> <LATEX> \mathrm{f}($)     # instead of <LATEX> \$
to_math(a$b)
#> <LATEX> \mathrm{$}(a b)  # instead of <LATEX> \mathrm{\$}(a b)
to_math(`\\`)
#> <LATEX> \mathrm{f}(\)     # instead of <LATEX> \\
to_math(`%`)
#> <LATEX> \mathrm{f}(%)     # instead of <LATEX> \%
```

To adjust this behaviour, we need an escape function with methods for the `character` and `advr_latex` classes.

(Note that we must first repeat the underlying code from *Advanced R*. However, since this would be a bit verbose, and not very meaningful, we will not show this step here.)

```
escape_latex <- function(x) UseMethod("escape_latex")

escape_latex.character <- function(x) {
  x <- gsub("^\\\\$", "\\\\\\\\", x)
  x <- gsub("^\\$$", "\\\\$", x)
  x <- gsub("^\\%$", "\\\\%", x)

  latex(x)
}

escape_latex.advr_latex <- function(x) x
```

We apply `escape_latex()` within `latex_env()` when creating environments for unknown symbols and unknown functions. For the unknown function, we need to modify `unknown_op()` first.

```
unknown_op <- function(op) {
  new_function(
    exprs(... = ),
    expr({
      contents <- paste(..., collapse = ", ")
      paste0(
        !!paste0("\\mathrm{", escape_latex(op), "}("), contents, ")"
      )
    })
  )
}

latex_env <- function(expr) {
  calls <- all_calls(expr)
  call_list <- map(set_names(calls), unknown_op)
  call_env <- as_environment(call_list)

  # Known functions
  f_env <- env_clone(f_env, call_env)

  # Default symbols
  names <- all_names(expr)
  symbol_env <- as_environment(set_names(escape_latex(names), names),
                               parent = f_env)

  # Known symbols
  greek_env <- env_clone(greek_env, parent = symbol_env)
```

```
  greek_env
}
```

Now, we can validate to_math() on the test cases from above.

```
to_math(`$`)
#> <LATEX> \$
to_math(a$b)
#> <LATEX> \mathrm{\$}(a b)
to_math(`\\`)
#> <LATEX> \\
to_math(`%`)
#> <LATEX> \%
```

Q2: Complete the DSL to support all the functions that plotmath supports.

A: You can see all supported functions in ?plotmath. There are a lot (!) so here we choose to implement a representative sample:

```
to_math(x %+-% y)
to_math(x %*% y)
to_math(x %->% y)
to_math(bold(x))
to_math(x != y)
```

Implementing the rest is just a mechanical application of the same principles with more LaTex expressions, which can be found on Wikipedia (https://en.wikipedia.org/wiki/Help:Displaying_a_formula).

To provide these translations, we'll follow the LaTeX section from *Advanced R* from the beginning. This makes it easier to keep an overview, as we just need to insert the specific changes at the relevant parts.

Let's start and repeat the converter function to_math() from the textbook.

```
to_math <- function(x) {
  expr <- enexpr(x)
  out <- eval_bare(expr, latex_env(expr))

  latex(out)
}

latex <- function(x) structure(x, class = "advr_latex")
print.advr_latex <- function(x) {
```

```
  cat("<LATEX> ", x, "\n", sep = "")
}
```

One specific property in this setting is that the environment where `to_math()` evaluates the expression is not constant, but depends on what we already know about the expression.

Next, we start building up `latex_env()`, which contains a chain of all the necessary environments which `to_math()` checks to evaluate the expression in.

The first environment is the one for Greek letters.

```
greek <- c(
  "alpha", "theta", "tau", "beta", "vartheta", "pi", "upsilon",
  "gamma", "varpi", "phi", "delta", "kappa", "rho",
  "varphi", "epsilon", "lambda", "varrho", "chi", "varepsilon",
  "mu", "sigma", "psi", "zeta", "nu", "varsigma", "omega", "eta",
  "xi", "Gamma", "Lambda", "Sigma", "Psi", "Delta", "Xi",
  "Upsilon", "Omega", "Theta", "Pi", "Phi"
)
greek_list <- set_names(paste0("\\", greek), greek)
greek_env <- as_environment(greek_list)

latex_env <- function(expr) {
  greek_env
}
```

We already know from *Advanced R* that e.g. `to_math(pi)` now correctly converts to `\\pi`. So, let's move on to the next one.

Here, it'll become a bit more technical. Not every symbol is Greek (and not every part of an expression is a symbol). To find out which symbols are present within the expression, first, we use an approach from section 5 of the expressions chapter (`https://adv-r.hadley.nz/expressions.html#ast-funs`) (walking the AST to find all symbols) where Hadley recursively walks the AST to distinguish between different expression element types.

Let's briefly repeat the helpers defined in that section:

```
expr_type <- function(x) {
  if (rlang::is_syntactic_literal(x)) {
    "constant"
  } else if (is.symbol(x)) {
    "symbol"
  } else if (is.call(x)) {
```

```
    "call"
  } else if (is.pairlist(x)) {
    "pairlist"
  } else {
    typeof(x)
  }
}

switch_expr <- function(x, ...) {
  switch(expr_type(x),
         ...,
         stop("Don't know how to handle type ",
              typeof(x), call. = FALSE)
  )
}

flat_map_chr <- function(.x, .f, ...) {
  purrr::flatten_chr(purrr::map(.x, .f, ...))
}
```

This lets us define all_names(), which returns the desired symbols, already converted to characters.

```
all_names_rec <- function(x) {
  switch_expr(x,
              constant = character(),
              symbol =   as.character(x),
              call =     flat_map_chr(as.list(x[-1]), all_names)
  )
}

all_names <- function(x) {
  unique(all_names_rec(x))
}

all_names(expr(x + y + f(a, b, c, 10)))
#> [1] "x" "y" "a" "b" "c"
```

We use all_names() now within latex_env() to create an environment of the symbols which were found within the expression. This environment will be set as the parent environment of greek_env.

```
latex_env <- function(expr) {
  # Unknown symbols
  names <- all_names(expr)
  symbol_env <- as_environment(set_names(names))

  # Known symbols
  env_clone(greek_env, parent = symbol_env)
}
```

In this way, `to_math()` will first convert all known Greek letters (found in `greek_env`) and then any other symbols, which are left as is (in this implementation).

We also have to add support for functions. This will give us the opportunity to insert some specific support for `plotmath` functions.

To support a whole bunch of unary and binary functions within the function environment (`f_env`), which will be added next to `latex_env`, Hadley defines the following two helpers in *Advanced R*.

```
unary_op <- function(left, right) {
  new_function(
    exprs(e1 = ),
    expr(
      paste0(!!left, e1, !!right)
    ),
    caller_env()
  )
}

binary_op <- function(sep) {
  new_function(
    exprs(e1 = , e2 = ),
    expr(
      paste0(e1, !!sep, e2)
    ),
    caller_env()
  )
}
```

While defining the function environment, `f_env`, we mostly continue to copy the exact code from *Advanced R*. However, at the bottom we add a short section where we define some extra conversions which are part of `plotmath` (and selected above in our intro to this solution).

```
f_env <- child_env(
  # Binary operators
  .parent = empty_env(),
  `+` = binary_op(" + "),
  `-` = binary_op(" - "),
  `*` = binary_op(" * "),
  `/` = binary_op(" / "),
  `^` = binary_op("^"),
  `[` = binary_op("_"),

  # Grouping
  `{` = unary_op("\\left{ ", " \\right}"),
  `(` = unary_op("\\left( ", " \\right)"),
  paste = paste,

  # Other math functions
  sqrt = unary_op("\\sqrt{", "}"),
  sin =  unary_op("\\sin(", ")"),
  log =  unary_op("\\log(", ")"),
  abs =  unary_op("\\left| ", "\\right| "),
  frac = function(a, b) {
    paste0("\\frac{", a, "}{", b, "}")
  },

  # Labelling
  hat =   unary_op("\\hat{", "}"),
  tilde = unary_op("\\tilde{", "}"),

  # Plotmath
  `%+-%` = binary_op(" \\pm "),
  `%*%`  = binary_op(" \\times "),
  `%->%` = binary_op(" \\rightarrow "),
  bold = unary_op("\\textbf{", "}"),
  `!=` = binary_op(" \\neq ")
)
```

Again we extend `latex_env()` to include the additional environment, `f_env`, which must be the parent of the symbol environment (which is the parent of the Greek symbol environment).

```
latex_env <- function(expr) {
  # Known functions
  f_env
```

```
  # Default symbols
  names <- all_names(expr)
  symbol_env <- as_environment(set_names(names), parent = f_env)

  # Known symbols
  greek_env <- env_clone(greek_env, parent = symbol_env)

  greek_env
}
```

Now, we can finally check if our new functionality works:

```
# New plotmath functionality
to_math(x %+-% y)
#> <LATEX> x \pm y
to_math(x %*% y)
#> <LATEX> x \times y
to_math(x %->% y)
#> <LATEX> x \rightarrow y
to_math(bold(x))
#> <LATEX> \textbf{x}
to_math(x != y)
#> <LATEX> x \neq y

# Other examples from Advanced R
to_math(sin(x + pi))
#> <LATEX> \sin(x + \pi)
to_math(log(x[i]^2))
#> <LATEX> \log(x_i^2)
to_math(sin(sin))
#> <LATEX> \sin(sin)
```

If we wanted, we could include further `plotmath` functions in this step. If this collides with other functions at some point, we could just create our own `f_plotmath_env` to support more `plotmath` functions. In that case we would need to add this environment also to `latex_env()` (as child environment of the function environment `f_env`).

To complete this answer, we will also add the support for unknown functions. Similarly as for the unknown symbols, we use the trick to recursively run the AST and just reuse the code from *Advanced R*.

```
all_calls_rec <- function(x) {
  switch_expr(x,
              constant = ,
              symbol =   character(),
              call = {
                fname <- as.character(x[[1]])
                children <- flat_map_chr(as.list(x[-1]), all_calls)
                c(fname, children)
              }
  )
}

all_calls <- function(x) {
  unique(all_calls_rec(x))
}

all_calls(expr(f(g + b, c, d(a))))
#> [1] "f" "+" "d"

unknown_op <- function(op) {
  new_function(
    exprs(... = ),
    expr({
      contents <- paste(..., collapse = ", ")
      paste0(!!paste0("\\mathrm{", op, "}("), contents, ")")
    })
  )
}
```

Of course, we need to add the new `call_env` also to `latex_env()`.

```
latex_env <- function(expr) {
  calls <- all_calls(expr)
  call_list <- map(set_names(calls), unknown_op)
  call_env <- as_environment(call_list)

  # Known functions
  f_env <- env_clone(f_env, call_env)

  # Default symbols
  names <- all_names(expr)
  symbol_env <- as_environment(set_names(names), parent = f_env)
```

```
  # Known symbols
  greek_env <- env_clone(greek_env, parent = symbol_env)
  greek_env
}
```

Finally, we rerun our tests and double check the newly supported `plotmath` operators.

```
# New plotmath functionality
to_math(x %+-% y)
#> <LATEX> x \pm y
to_math(x %*% y)
#> <LATEX> x \times y
to_math(x %->% y)
#> <LATEX> x \rightarrow y
to_math(bold(x))
#> <LATEX> \textbf{x}
to_math(x != y)
#> <LATEX> x \neq y

# Other examples from Advanced R
to_math(sin(x + pi))
#> <LATEX> \sin(x + \pi)
to_math(log(x[i]^2))
#> <LATEX> \log(x_i^2)
to_math(sin(sin))
#> <LATEX> \sin(sin)

# Unknown functions
to_math(f(g(x)))
#> <LATEX> \mathrm{f}(\mathrm{g}(x))
```

Part V

Techniques

23

Measuring performance

23.2 Profiling

Q1: Profile the following function with `torture = TRUE`. What is surprising? Read the source code of `rm()` to figure out what's going on.

```
f <- function(n = 1e5) {
  x <- rep(1, n)
  rm(x)
}
```

A: We expect `f()` to create a vector (`x`) of length `n`, which is then removed so that `f()` just returns `NULL`. When we profile this function, it executes too fast for meaningful results.

```
profvis::profvis(f())
#> Error in parse_rprof(prof_output, expr_source): No parsing data
#> available. Maybe your function was too fast?
```

Setting `torture = TRUE` triggers garbage collection after every memory allocation call, which may be useful for more exact memory profiling.

```
profvis::profvis(f(), torture = TRUE)
```

Surprisingly, profiling `f()` like this takes a very long time. What could be the reason?

We follow the hint in the question and inspect the source code of `rm()`:

```
function (..., list = character(), pos = -1,
          envir = as.environment(pos),
          inherits = FALSE)
{
```

DOI: 10.1201/9781003175414-23

```
  dots <- match.call(expand.dots = FALSE)$...
  if (
    length(dots) && !all(
      vapply(dots, function(x) is.symbol(x) ||
              is.character(x), NA, USE.NAMES = FALSE)
    )
  )
    stop("... must contain names or character strings")
  names <- vapply(dots, as.character, "")
  if (length(names) == 0L)
    names <- character()
  list <- .Primitive("c")(list, names)
  .Internal(remove(list, envir, inherits))
}
```

`rm()` does a surprising amount of work to get the name of the object to delete because it relies on non-standard evaluation.

We can make the job of `rm()` considerably simpler by using the `list` argument:

```
f2 <- function(n = 1e5) {
  x <- rep(1, n)
  rm(list = "x")
}
profvis::profvis(f2(), torture = TRUE)
```

Unfortunately, this still takes too long, and we are literally stuck in profiling.

Anecdotally, one of the authors once finished the profiling under an older R version. But the output seemed to be not very meaningful.

In conclusion, this question appears to be unanswerable for us, even for Hadley.

23.3 Microbenchmarking

Q1: Instead of using `bench::mark()`, you could use the built-in function `system.time()`. But `system.time()` is much less precise, so you'll need to repeat each operation many times with a loop, and then divide to find the average time of each operation, as in the code below.

```r
n <- 1e6
system.time(for (i in 1:n) sqrt(x)) / n
system.time(for (i in 1:n) x ^ 0.5) / n
```

How do the estimates from `system.time()` compare to those from `bench::mark()`? Why are they different?

A: We first microbenchmark these two expressions using `bench::mark()` [Hester, 2020] and observe that the mean is not reported (as it is generally more affected by outliers).

```r
n <- 1e6
x <- runif(100)

bench_df <- bench::mark(
  sqrt(x),
  x ^ 0.5,
  iterations = n
)

bench_df
#> # A tibble: 2 x 6
#>   expression      min   median `itr/sec` mem_alloc `gc/sec`
#>   <bch:expr> <bch:tm> <bch:tm>     <dbl> <bch:byt>    <dbl>
#> 1 sqrt(x)    272.99ns    322ns  1795203.      848B     30.5
#> 2 x^0.5        5.05µs   5.49µs   174914.      848B      6.12
```

We need to access the raw data, so we can compare the results of both benchmarking approaches.

```r
t1_bench <- mean(unlist(bench_df[1, "time"]))
t2_bench <- mean(unlist(bench_df[2, "time"]))

t1_systime <- system.time(for (i in 1:n) sqrt(x)) / n
t2_systime <- system.time(for (i in 1:n) x ^ 0.5) / n
```

We see, that both approaches get the order of magnitude right. We assume, that the `bench::mark()`-results may be a little more accurate, because of its high precision timer. There may also be overhead introduced by the for loop in the `system.time()`-approach.

```r
# Compare the results
t1_systime["elapsed"]
```

```
#>  elapsed
#> 5.88e-07
t1_bench
#> [1] 6.75e-07

t2_systime["elapsed"]
#>  elapsed
#> 5.36e-06
t2_bench
#> [1] 5.83e-06
```

Side Note: take a look at `?proc.time` if you want to learn about the differences between "user", "system" and "elapsed" time.

Q2: Here are two other ways to compute the square root of a vector. Which do you think will be fastest? Which will be slowest? Use microbenchmarking to test your answers.

```
x ^ (1 / 2)
exp(log(x) / 2)
```

A: To compare these approaches, we'll `bench::mark()` them and sort the result by the median execution time.

```
x <- runif(100)

bm <- bench::mark(
  sqrt(x),
  x^0.5,
  x^(1 / 2),
  exp(log(x) / 2)
)

bm[order(bm$median), ]
#> # A tibble: 4 x 6
#>   expression          min   median `itr/sec` mem_alloc `gc/sec`
#>   <bch:expr>     <bch:tm> <bch:tm>     <dbl> <bch:byt>    <dbl>
#> 1 sqrt(x)        312.11ns 394.07ns  2029264.      848B     203.
#> 2 exp(log(x)/2)    3.12µs   3.26µs   287368.      848B       0
#> 3 x^(1/2)          5.67µs   5.78µs   163139.      848B       0
#> 4 x^0.5            5.57µs   6.95µs   142455.      848B       0
```

As one might expect the idiomatic primitive function `sqrt()` is the fastest. The approach `exp(log(x) / 2)` which builds on two other primitive functions is second, even though already considerably slower. The other calculations are even slower: `x ^ 0.5` is faster than `x ^ (1 / 2)`, because `0.5` requires less computation than `(1 / 2)`.

24

Improving performance

24.3 Checking for existing solutions

Q1: What are faster alternatives to `lm`? Which are specifically designed to work with larger datasets?

A: The CRAN task view for high-performance computing (`https://cran.rstudio.com/web/views/HighPerformanceComputing.html`) provides many recommendations. For this question, we are most interested in the section on "Large memory and out-of-memory data". We could for example give `biglm::biglm()` [Lumley, 2020], `speedglm::speedlm()` [Enea, 2017] or `RcppEigen::fastLm()` [Bates and Eddelbuettel, 2013] a try.

For small datasets, we observe only minor performance gains (or even a small cost):

```
penguins <- palmerpenguins::penguins

bench::mark(
  "lm" = lm(
    body_mass_g ~ bill_length_mm + species, data = penguins
  ) %>% coef(),
  "biglm" = biglm::biglm(
    body_mass_g ~ bill_length_mm + species, data = penguins
  ) %>% coef(),
  "speedglm" = speedglm::speedlm(
    body_mass_g ~ bill_length_mm + species, data = penguins
  ) %>% coef(),
  "fastLm" = RcppEigen::fastLm(
    body_mass_g ~ bill_length_mm + species, data = penguins
  ) %>% coef()
)
#> # A tibble: 4 x 6
#>   expression      min   median `itr/sec` mem_alloc `gc/sec`
#>   <bch:expr> <bch:tm> <bch:tm>     <dbl> <bch:byt>    <dbl>
```

DOI: 10.1201/9781003175414-24

```
#> 1 lm            773µs    838µs     1151.  961.51KB     4.09
#> 2 biglm         598µs    618µs     1563.    5.39MB     4.07
#> 3 speedglm      954µs    976µs      990.   62.33MB     4.07
#> 4 fastLm        619µs    645µs     1490.    3.42MB     6.37
```

For larger datasets the selection of the appropriate method is of greater relevance:

```
eps <- rnorm(100000)
x1 <- rnorm(100000, 5, 3)
x2 <- rep(c("a", "b"), 50000)
y <- 7 * x1 + (x2 == "a") + eps
td <- data.frame(y = y, x1 = x1, x2 = x2, eps = eps)

bench::mark(
  "lm" = lm(y ~ x1 + x2, data = td) %>% coef(),
  "biglm" = biglm::biglm(y ~ x1 + x2, data = td) %>% coef(),
  "speedglm" = speedglm::speedlm(y ~ x1 + x2, data = td) %>% coef(),
  "fastLm" = RcppEigen::fastLm(y ~ x1 + x2, data = td) %>% coef()
)
#> # A tibble: 4 x 6
#>   expression      min   median `itr/sec` mem_alloc `gc/sec`
#>   <bch:expr> <bch:tm> <bch:tm>     <dbl> <bch:byt>    <dbl>
#> 1 lm           67.5ms   67.5ms      14.8      27MB    119.
#> 2 biglm        27.2ms   32.8ms      30.9    22.2MB     51.5
#> 3 speedglm     24.3ms     25ms      36.3    20.4MB     45.4
#> 4 fastLm       52.9ms   52.9ms      18.9    30.2MB    170.
```

For further speed improvements, you could install a linear algebra library optimised for your system (see `?speedglm::speedlm`).

> The functions of class 'speedlm' may speed up the fitting of LMs to large datasets. High performances can be obtained especially if R is linked against an optimized BLAS, such as ATLAS.

Tip: In case your dataset is stored in a database, you might want to check out the `{modeldb}` package (`https://github.com/tidymodels/modeldb`) [Kuhn, 2020] which executes the linear model code in the corresponding database backend.

Q2: What package implements a version of `match()` that's faster for repeated lookups? How much faster is it?

A: A web search points us to the `{fastmatch}` package [Urbanek, 2017]. We compare it to `base::match()` and observe an impressive performance gain.

```
table <- 1:100000
x <- sample(table, 10000, replace = TRUE)

bench::mark(
  match = match(x, table),
  fastmatch = fastmatch::fmatch(x, table)
)
#> # A tibble: 2 x 6
#>   expression      min   median `itr/sec` mem_alloc `gc/sec`
#>   <bch:expr> <bch:tm> <bch:tm>     <dbl> <bch:byt>    <dbl>
#> 1 match        15.5ms   16.2ms      60.1    1.46MB     2.07
#> 2 fastmatch   403.9µs  427.1µs    2259.   442.69KB     2.01
```

Q3: List four functions (not just those in base R) that convert a string into a date time object. What are their strengths and weaknesses?

A: The usual base R way is to use the `as.POSIXct()` generic and create a date time object of class `POSIXct` and type integer.

```
date_ct <- as.POSIXct("2020-01-01 12:30:25")
date_ct
#> [1] "2020-01-01 12:30:25 CET"
```

Under the hood `as.POSIXct()` employs `as.POSIXlt()` for the character conversion. This creates a date time object of class `POSIXlt` and type `list`.

```
date_lt <- as.POSIXlt("2020-01-01 12:30:25")
date_lt
#> [1] "2020-01-01 12:30:25 CET"
```

The `POSIXlt` class has the advantage that it carries the individual time components as attributes. This allows to extract the time components via typical list operators.

```
attributes(date_lt)
#> $names
#>  [1] "sec"    "min"    "hour"   "mday"   "mon"    "year"   "wday"
#>  [8] "yday"   "isdst"  "zone"   "gmtoff"
#>
#> $class
#> [1] "POSIXlt" "POSIXt"
date_lt$sec
#> [1] 25
```

However, while lists may be practical, basic calculations are often faster and require less memory for objects with underlying integer type.

```
date_lt2 <- rep(date_lt, 10000)
date_ct2 <- rep(date_ct, 10000)

bench::mark(
  date_lt2 - date_lt2,
  date_ct2 - date_ct2,
  date_ct2 - date_lt2
)
#> # A tibble: 3 x 6
#>   expression               min   median `itr/sec` mem_alloc `gc/sec`
#>   <bch:expr>          <bch:tm> <bch:tm>     <dbl> <bch:byt>    <dbl>
#> 1 date_lt2 - date_lt2   24.2ms   24.8ms      39.1    1.36MB     2.06
#> 2 date_ct2 - date_ct2   52.3µs   71.4µs   13382.   195.45KB    54.3
#> 3 date_ct2 - date_lt2   11.9ms   12.4ms      80.0  781.95KB     0
```

as.POSIXlt() in turn uses strptime() under the hood, which creates a similar date time object.

```
date_str <- strptime("2020-01-01 12:30:25",
                     format = "%Y-%m-%d %H:%M:%S")
identical(date_lt, date_str)
#> [1] TRUE
```

as.POSIXct() and as.POSIXlt() accept different character inputs by default (e.g. "2001-01-01 12:30" or "2001/1/1 12:30"). strptime() requires the format argument to be set explicitly, and provides a performance improvement in return.

```
bench::mark(
  as.POSIXct = as.POSIXct("2020-01-01 12:30:25"),
  as.POSIXct_format = as.POSIXct("2020-01-01 12:30:25",
    format = "%Y-%m-%d %H:%M:%S"
  ),
  strptime_fomat = strptime("2020-01-01 12:30:25",
    format = "%Y-%m-%d %H:%M:%S"
  )
)[1:3]
#> # A tibble: 3 x 3
#>   expression             min   median
#>   <bch:expr>        <bch:tm> <bch:tm>
#> 1 as.POSIXct          41.9µs  45.67µs
```

```
#> 2 as.POSIXct_format    21.7µs  23.15µs
#> 3 strptime_fomat        6.9µs   7.59µs
```

A fourth way is to use the converter functions from the `{lubridate}` package [Grolemund and Wickham, 2011], which contains wrapper functions (for the POSIXct approach) with an intuitive syntax. (There is a slight decrease in performance though.)

```
library(lubridate)
ymd_hms("2013-07-24 23:55:26")
#> [1] "2013-07-24 23:55:26 UTC"

bench::mark(
  as.POSIXct = as.POSIXct("2013-07-24 23:55:26", tz = "UTC"),
  ymd_hms = ymd_hms("2013-07-24 23:55:26")
)[1:3]
#> # A tibble: 2 x 3
#>   expression      min   median
#>   <bch:expr> <bch:tm> <bch:tm>
#> 1 as.POSIXct   40.6µs  49.33µs
#> 2 ymd_hms      2.38ms   2.77ms
```

For additional ways to convert characters into date time objects, have a look at the `{chron}`, the `{anytime}` and the `{fasttime}` packages. The `{chron}` package [James and Hornik, 2020] introduces new classes and stores times as fractions of days in the underlying double type, while it doesn't deal with time zones and daylight savings. The `{anytime}` package [Eddelbuettel, 2020] aims to convert "Anything to POSIXct or Date". The `{fasttime}` package [Urbanek, 2016] contains only one function, `fastPOSIXct()`.

Q4: Which packages provide the ability to compute a rolling mean?

A: A rolling mean is a useful statistic to smooth time-series, spatial and other types of data. The size of the rolling window usually determines the amount of smoothing and the number of missing values at the boundaries of the data.

The general functionality can be found in multiple packages, which vary in speed and flexibility of the computations. Here is a benchmark for several functions that all serve our purpose.

```
x <- 1:10
slider::slide_dbl(x, mean, .before = 1, .complete = TRUE)
#>  [1]  NA 1.5 2.5 3.5 4.5 5.5 6.5 7.5 8.5 9.5
```

```r
bench::mark(
  caTools = caTools::runmean(x, k = 2, endrule = "NA"),
  data.table = data.table::frollmean(x, 2),
  RcppRoll = RcppRoll::roll_mean(x, n = 2, fill = NA,
                                 align = "right"),
  slider = slider::slide_dbl(x, mean, .before = 1, .complete = TRUE),
  TTR = TTR::SMA(x, 2),
  zoo_apply = zoo::rollapply(x, 2, mean, fill = NA, align = "right"),
  zoo_rollmean = zoo::rollmean(x, 2, fill = NA, align = "right")
)
#> # A tibble: 7 x 6
#>   expression        min   median `itr/sec` mem_alloc `gc/sec`
#>   <bch:expr>   <bch:tm> <bch:tm>     <dbl> <bch:byt>    <dbl>
#> 1 caTools        76.5µs   93.6µs    10516.  165.69KB     20.2
#> 2 data.table     45.4µs   54.4µs    18221.    1.59MB     21.4
#> 3 RcppRoll       40.3µs   47.6µs    20828.   58.74KB     21.4
#> 4 slider         75.7µs   89.1µs    11037.        0B     21.3
#> 5 TTR           387.1µs  408.2µs     2340.    1.98MB      8.20
#> 6 zoo_apply     440.2µs  452.6µs     2150.  581.62KB     19.1
#> 7 zoo_rollmean    388µs  398.5µs     2430.    6.42KB     19.2
```

You may also take a look at an extensive example in the first edition of *Advanced R* (`http://adv-r.had.co.nz/Functionals.html`), which demonstrates how a rolling mean function can be created.

Q5: What are the alternatives to `optim()`?

A: According to its description (see `?optim`) `optim()` implements:

> General-purpose optimization based on Nelder–Mead, quasi-Newton and conjugate-gradient algorithms. It includes an option for box-constrained optimization and simulated annealing.

`optim()` allows to optimise a function (`fn`) on an interval with a specific method (`method = c("Nelder-Mead", "BFGS", "CG", "L-BFGS-B", "SANN", "Brent")`). Many detailed examples are given in the documentation. In the simplest case, we give `optim()` the starting value `par = 0` to calculate the minimum of a quadratic polynomial:

```r
optim(0, function(x) x^2 - 100 * x + 50,
  method = "Brent",
  lower = -1e20, upper = 1e20
)
#> $par
#> [1] 50
```

```
#>
#> $value
#> [1] -2450
#>
#> $counts
#> function gradient
#>       NA       NA
#>
#> $convergence
#> [1] 0
#>
#> $message
#> NULL
```

Since this solves a one-dimensional optimisation task, we could have also used `stats::optimize()`.

```
optimize(function(x) x^2 - 100 * x + 50, c(-1e20, 1e20))
#> $minimum
#> [1] 50
#>
#> $objective
#> [1] -2450
```

For more general alternatives, the appropriate choice highly depends on the type of optimisation you intend to do. The CRAN task view on optimisation and mathematical modelling (`https://cran.r-project.org/web/views/Optimization.html`) can serve as a useful reference. Here are a couple of examples:

- `{optimx}` [Nash and Varadhan, 2011, Nash, 2014] extends the `optim()` function with the same syntax but more `method` choices.
- `{RcppNumerical}` [Qiu et al., 2019] wraps several open source libraries for numerical computing (written in C++) and integrates them with R via `{Rcpp}`.
- `{DEoptim}` [Mullen et al., 2011] provides a global optimiser based on the Differential Evolution algorithm.

24.4 Doing as little as possible

Q1: What's the difference between `rowSums()` and `.rowSums()`?

A: When we inspect the source code of the user-facing `rowSums()`, we see that it is designed as a wrapper around `.rowSums()` with some input validation, conversions and handling of complex numbers.

```
rowSums
#> function (x, na.rm = FALSE, dims = 1L)
#> {
#>     if (is.data.frame(x))
#>         x <- as.matrix(x)
#>     if (!is.array(x) || length(dn <- dim(x)) < 2L)
#>         stop("'x' must be an array of at least two dimensions")
#>     if (dims < 1L || dims > length(dn) - 1L)
#>         stop("invalid 'dims'")
#>     p <- prod(dn[-(id <- seq_len(dims))])
#>     dn <- dn[id]
#>     z <- if (is.complex(x))
#>         .Internal(rowSums(Re(x), prod(dn), p, na.rm)) + (0+1i) *
#>             .Internal(rowSums(Im(x), prod(dn), p, na.rm))
#>     else .Internal(rowSums(x, prod(dn), p, na.rm))
#>     if (length(dn) > 1L) {
#>         dim(z) <- dn
#>         dimnames(z) <- dimnames(x)[id]
#>     }
#>     else names(z) <- dimnames(x)[[1L]]
#>     z
#> }
#> <bytecode: 0x5627c327f338>
#> <environment: namespace:base>
```

`.rowSums()` calls an internal function, which is built into the R interpreter. These compiled functions can be very fast.

```
.rowSums
#> function (x, m, n, na.rm = FALSE)
#> .Internal(rowSums(x, m, n, na.rm))
#> <bytecode: 0x5627c2a43090>
#> <environment: namespace:base>
```

However, as our benchmark reveals almost identical computing times, we prefer the safer variant over the internal function for this case.

```
m <- matrix(rnorm(1e6), nrow = 1000)
```

```
bench::mark(
  rowSums(m),
  .rowSums(m, 1000, 1000)
)
#> # A tibble: 2 x 6
#>   expression                   min median `itr/sec` mem_alloc `gc/sec`
#>   <bch:expr>              <bch:t> <bch:>     <dbl> <bch:byt>    <dbl>
#> 1 rowSums(m)               2.21ms 2.25ms      427.    7.86KB        0
#> 2 .rowSums(m, 1000, 1000)  2.15ms 2.29ms      430.    7.86KB        0
```

Q2: Make a faster version of `chisq.test()` that only computes the chi-square test statistic when the input is two numeric vectors with no missing values. You can try simplifying `chisq.test()` or by coding from the mathematical definition (`http://en.wikipedia.org/wiki/Pearson%27s_chi-squared_test`).

A: We aim to speed up our reimplementation of `chisq.test()` by *doing less*.

```
chisq.test2 <- function(x, y) {
  m <- rbind(x, y)
  margin1 <- rowSums(m)
  margin2 <- colSums(m)
  n <- sum(m)
  me <- tcrossprod(margin1, margin2) / n

  x_stat <- sum((m - me)^2 / me)
  df <- (length(margin1) - 1) * (length(margin2) - 1)
  p.value <- pchisq(x_stat, df = df, lower.tail = FALSE)

  list(x_stat = x_stat, df = df, p.value = p.value)
}
```

We check if our new implementation returns the same results and benchmark it afterwards.

```
a <- 21:25
b <- seq(21, 29, 2)
m <- cbind(a, b)

chisq.test(m) %>% print(digits=5)
#>
#>  Pearson's Chi-squared test
#>
#> data:  m
```

```
#> X-squared = 0.162, df = 4, p-value = 1
chisq.test2(a, b)
#> $x_stat
#> [1] 0.162
#>
#> $df
#> [1] 4
#>
#> $p.value
#> [1] 0.997
```

```
bench::mark(
  chisq.test(m),
  chisq.test2(a, b),
  check = FALSE
)
#> # A tibble: 2 x 6
#>   expression              min   median `itr/sec` mem_alloc `gc/sec`
#>   <bch:expr>         <bch:tm> <bch:tm>     <dbl> <bch:byt>    <dbl>
#> 1 chisq.test(m)        58.1µs   67.4µs    14036.        0B     4.13
#> 2 chisq.test2(a, b)    17.3µs   18.8µs    52045.        0B     5.21
```

Q3: Can you make a faster version of `table()` for the case of an input of two integer vectors with no missing values? Can you use it to speed up your chi-square test?

A: When analysing the source code of `table()` we aim to omit everything unnecessary and extract the main building blocks. We observe that `table()` is powered by `tabulate()` which is a very fast counting function. This leaves us with the challenge to compute the pre-processing as performant as possible.

First, we calculate the dimensions and names of the output table. Then we use `fastmatch::fmatch()` to map the elements of each vector to their position within the vector itself (i.e. the smallest value is mapped to `1L`, the second smallest value to `2L`, etc.). Following the logic within `table()` we combine and shift these values to create a mapping of the integer pairs in our data to the index of the output table. After applying these lookups `tabulate()` counts the values and returns an integer vector with counts for each position in the table. As a last step, we reuse the code from `table()` to assign the correct dimension and class.

```
table2 <- function(a, b){

  a_s <- sort(unique(a))
```

```
  b_s <- sort(unique(b))

  a_l <- length(a_s)
  b_l <- length(b_s)

  dims <- c(a_l, b_l)
  pr <- a_l * b_l
  dn <- list(a = a_s, b = b_s)

  bin <- fastmatch::fmatch(a, a_s) +
    a_l * fastmatch::fmatch(b, b_s) - a_l
  y <- tabulate(bin, pr)

  y <- array(y, dim = dims, dimnames = dn)
  class(y) <- "table"

  y
}

a <- sample(100, 10000, TRUE)
b <- sample(100, 10000, TRUE)

bench::mark(
  table(a, b),
  table2(a, b)
)
#> # A tibble: 2 x 6
#>   expression        min   median `itr/sec` mem_alloc `gc/sec`
#>   <bch:expr>   <bch:tm> <bch:tm>     <dbl> <bch:byt>    <dbl>
#> 1 table(a, b)    1.03ms   1.23ms      815.    1.29MB     22.2
#> 2 table2(a, b) 394.66µs 450.27µs     2144.  694.56KB     19.8
```

Since we didn't use `table()` in our `chisq.test2()`-implementation, we cannot benefit from the slight performance gain from `table2()`.

24.5 Vectorise

Q1: The density functions, e.g. `dnorm()`, have a common interface. Which arguments are vectorised over? What does `rnorm(10, mean = 10:1)` do?

A: We can get an overview of the interface of these functions via `?dnorm`:

```
dnorm(x, mean = 0, sd = 1, log = FALSE)
pnorm(q, mean = 0, sd = 1, lower.tail = TRUE, log.p = FALSE)
qnorm(p, mean = 0, sd = 1, lower.tail = TRUE, log.p = FALSE)
rnorm(n, mean = 0, sd = 1)
```

These functions are vectorised over their numeric arguments, which includes the first argument (`x`, `q`, `p`, `n`) as well as `mean` and `sd`.

`rnorm(10, mean = 10:1)` generates ten random numbers from different normal distributions. These normal distributions differ in their means. The first has mean 10, the second mean 9, the third mean 8 and so on.

Q2: Compare the speed of `apply(x, 1, sum)` with `rowSums(x)` for varying sizes of `x`.

A: We compare the two functions for square matrices of increasing size:

```
rowsums <- bench::press(
  p = seq(500, 5000, length.out = 10),
  {
    mat <- tcrossprod(rnorm(p), rnorm(p))
    bench::mark(
      rowSums = rowSums(mat),
      apply = apply(mat, 1, sum)
    )
  }
)
#> Running with:
#>        p
#>  1   500
#>  2  1000
#>  3  1500
#>  4  2000
#>  5  2500
#>  6  3000
#>  7  3500
#>  8  4000
#>  9  4500
#> 10  5000

library(ggplot2)

rowsums %>%
```

```
  summary() %>%
  dplyr::mutate(Approach = as.character(expression)) %>%
  ggplot(
    aes(p, median, color = Approach, group = Approach)) +
  geom_point() +
  geom_line() +
  labs(x = "Number of Rows and Columns",
       y = "Median (s)") +
  theme(legend.position = "top")
```

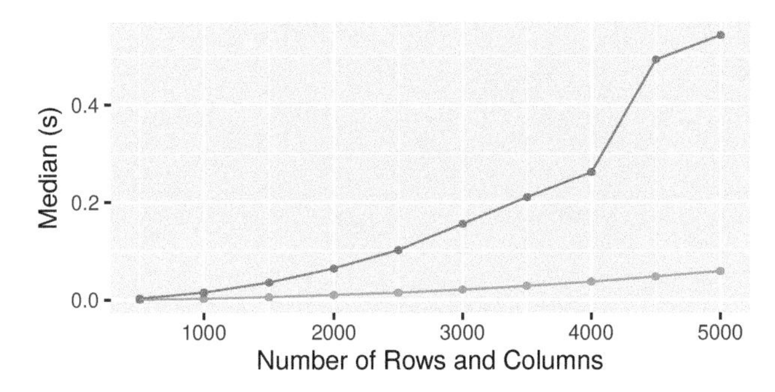

We can see that the difference in performance is negligible for small matrices but becomes more and more relevant as the size of the data increases. `apply()` is a very versatile tool, but it's not "vectorised for performance" and not as optimised as `rowSums()`.

Q3: How can you use `crossprod()` to compute a weighted sum? How much faster is it than the naive `sum(x * w)`?

A: We can hand the vectors to `crossprod()`, which converts them to row- and column-vectors and then multiplies these. The result is the dot product, which corresponds to a weighted sum.

```
x <- rnorm(10)
w <- rnorm(10)
all.equal(sum(x * w), crossprod(x, w)[[1]])
#> [1] TRUE
```

A benchmark of both approaches for different vector lengths indicates that the `crossprod()` variant is almost twice as fast as `sum(x * w)`.

```
weightedsum <- bench::press(
  n = 1:10,
```

```
  {
    x <- rnorm(n * 1e6)
    bench::mark(
      sum = sum(x * x),
      crossprod = crossprod(x, x)[[1]]
    )
  }
)
#> Running with:
#>        n
#>  1     1
#>  2     2
#>  3     3
#>  4     4
#>  5     5
#>  6     6
#>  7     7
#>  8     8
#>  9     9
#> 10    10

weightedsum %>%
  summary() %>%
  dplyr::mutate(Approach = as.character(expression)) %>%
  ggplot(aes(n, median, color = Approach, group = Approach)) +
  geom_point() +
  geom_line() +
  labs(x = "Vector length (millions)",
       y = "Median (s)") +
  theme(legend.position = "top")
```

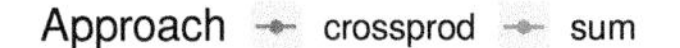

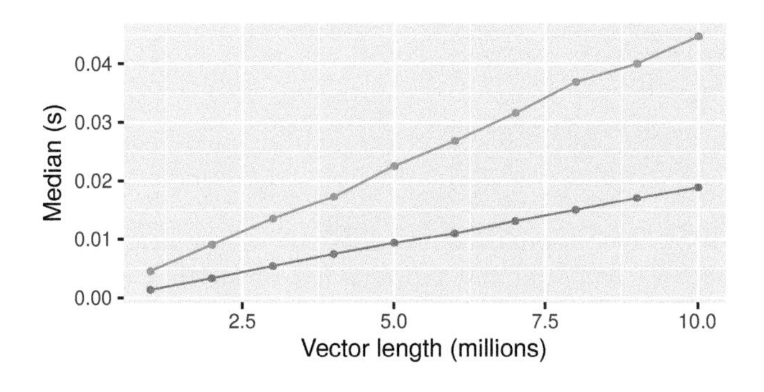

25

Rewriting R code in C++

25.2 Getting started with C++

Q1: With the basics of C++ in hand, it's now a great time to practice by reading and writing some simple C++ functions. For each of the following functions, read the code and figure out what the corresponding base R function is. You might not understand every part of the code yet, but you should be able to figure out the basics of what the function does.

```
double f1(NumericVector x) {
  int n = x.size();
  double y = 0;

  for(int i = 0; i < n; ++i) {
    y += x[i] / n;
  }
  return y;
}

NumericVector f2(NumericVector x) {
  int n = x.size();
  NumericVector out(n);

  out[0] = x[0];
  for(int i = 1; i < n; ++i) {
    out[i] = out[i - 1] + x[i];
  }
  return out;
}

bool f3(LogicalVector x) {
  int n = x.size();

  for(int i = 0; i < n; ++i) {
```

DOI: 10.1201/9781003175414-25

```
    if (x[i]) return true;
  }
  return false;
}

int f4(Function pred, List x) {
  int n = x.size();

  for(int i = 0; i < n; ++i) {
    LogicalVector res = pred(x[i]);
    if (res[0]) return i + 1;
  }
  return 0;
}

NumericVector f5(NumericVector x, NumericVector y) {
  int n = std::max(x.size(), y.size());
  NumericVector x1 = rep_len(x, n);
  NumericVector y1 = rep_len(y, n);

  NumericVector out(n);

  for (int i = 0; i < n; ++i) {
    out[i] = std::min(x1[i], y1[i]);
  }

  return out;
}
```

A: The code above corresponds to the following base R functions:

- f1: `mean()`
- f2: `cumsum()`
- f3: `any()`
- f4: `Position()`
- f5: `pmin()`

Q2: To practice your function writing skills, convert the following functions into C++. For now, assume the inputs have no missing values.

1. `all()`.
2. `cumprod()`, `cummin()`, `cummax()`.
3. `diff()`. Start by assuming lag 1, and then generalise for lag `n`.

4. range().
5. var(). Read about the approaches you can take on Wikipedia (http://en.wikipedia.org/wiki/Algorithms_for_calculating_variance). Whenever implementing a numerical algorithm, it's always good to check what is already known about the problem.

A: Let's port these functions to C++.

1. all()

```
bool allC(LogicalVector x) {
  int n = x.size();

  for (int i = 0; i < n; ++i) {
    if (!x[i]) return false;
  }
  return true;
}
```

2. cumprod(), cummin(), cummax().

```
NumericVector cumprodC(NumericVector x) {
  int n = x.size();
  NumericVector out(n);

  out[0] = x[0];
  for (int i = 1; i < n; ++i) {
    out[i]  = out[i - 1] * x[i];
  }
  return out;
}

NumericVector cumminC(NumericVector x) {
  int n = x.size();
  NumericVector out(n);

  out[0] = x[0];
  for (int i = 1; i < n; ++i) {
    out[i]  = std::min(out[i - 1], x[i]);
  }
  return out;
}
```

```
NumericVector cummaxC(NumericVector x) {
  int n = x.size();
  NumericVector out(n);

  out[0] = x[0];
  for (int i = 1; i < n; ++i) {
    out[i]  = std::max(out[i - 1], x[i]);
  }
  return out;
}
```

3. `diff()` (Start by assuming lag 1, and then generalise for lag `n`.)

```
NumericVector diffC(NumericVector x) {
  int n = x.size();
  NumericVector out(n - 1);

  for (int i = 1; i < n; i++) {
    out[i - 1] = x[i] - x[i - 1];
  }
  return out ;
}

NumericVector difflagC(NumericVector x, int lag = 1) {
  int n = x.size();

  if (lag >= n) stop("`lag` must be less than `length(x)`.");

  NumericVector out(n - lag);

  for (int i = lag; i < n; i++) {
    out[i - lag] = x[i] - x[i - lag];
  }
  return out;
}
```

4. `range()`

```
NumericVector rangeC(NumericVector x) {
  double omin = x[0], omax = x[0];
```

```
  int n = x.size();

  if (n == 0) stop("`length(x)` must be greater than 0.");

  for (int i = 1; i < n; i++) {
    omin = std::min(x[i], omin);
    omax = std::max(x[i], omax);
  }

  NumericVector out(2);
  out[0] = omin;
  out[1] = omax;
  return out;
}
```

5. var()

```
double varC(NumericVector x) {
  int n = x.size();

  if (n < 2) {
    return NA_REAL;
  }

  double mx = 0;
  for (int i = 0; i < n; ++i) {
    mx += x[i] / n;
  }

  double out = 0;
  for (int i = 0; i < n; ++i) {
    out += pow(x[i] - mx, 2);
  }

  return out / (n - 1);
}
```

25.4 Missing values

Q1: Rewrite any of the functions from the first exercise to deal with missing values. If `na.rm` is true, ignore the missing values. If `na.rm` is false, return a missing value if the input contains any missing values. Some good functions to practice with are `min()`, `max()`, `range()`, `mean()`, and `var()`.

A: For this exercise we start with `minC()` and extend it so it can deal with missing values. We introduce an `na_rm` argument to make `minC()` aware of `NA`s. In case `x` contains exclusively `NA` values `minC()` should return `Inf` for `na_rm = TRUE`. For the return values vector data types are used to avoid irregular type conversions.

```
#include <Rcpp.h>
using namespace Rcpp;

// [[Rcpp::export]]
NumericVector minC(NumericVector x, bool na_rm = false) {
  int n = x.size();
  NumericVector out = NumericVector::create(R_PosInf);

  if (na_rm) {
    for (int i = 0; i < n; ++i) {
      if (x[i] == NA_REAL) {
        continue;
      }
      if (x[i] < out[0]) {
        out[0] = x[i];
      }
    }
  } else {
    for (int i = 0; i < n; ++i) {
      if (NumericVector::is_na(x[i])) {
        out[0] = NA_REAL;
        return out;
      }
      if (x[i] < out[0]) {
        out[0] = x[i];
      }
    }
  }
```

```
  return out;
}
```

```
minC(c(2:4, NA))
#> [1] NA
minC(c(2:4, NA), na_rm = TRUE)
#> [1] 2
minC(c(NA, NA), na_rm = TRUE)
#> [1] Inf
```

We also extend `anyC()` so it can deal with missing values. Please note that this (again) introduces some code duplication. This could be avoided by moving the check for missing values to the inner loop at the expense of a slight decrease of performance. Here we use `LogicalVector` as return type. If we would use `bool` instead, the C++ `NA_LOGICAL` would be converted into R's logical `TRUE`.

```
#include <Rcpp.h>
using namespace Rcpp;

// [[Rcpp::export]]
LogicalVector anyC(LogicalVector x, bool na_rm = false) {
  int n = x.size();
  LogicalVector out = LogicalVector::create(false);

  if (na_rm == false) {
    for (int i = 0; i < n; ++i) {
      if (LogicalVector::is_na(x[i])) {
        out[0] = NA_LOGICAL;
        return out;
      } else {
        if (x[i]) {
          out[0] = true;
        }
      }
    }
  }

  if (na_rm) {
    for (int i = 0; i < n; ++i) {
      if (LogicalVector::is_na(x[i])) {
        continue;
      }
```

```
      if (x[i]) {
        out[0] = true;
        return out;
      }
    }
  }

  return out;
}
```

```
anyC(c(NA, TRUE))  # any(c(NA, TRUE)) would return TRUE in this case
#> [1] NA
anyC(c(NA, TRUE), na_rm = TRUE)
#> [1] TRUE
```

Q2: Rewrite `cumsum()` and `diff()` so they can handle missing values. Note that these functions have slightly more complicated behaviour.

A: Our `NA`-aware `cumsumC()` function will return a vector of the same length as `x`. By default (`na_rm = FALSE`) all values following the first `NA` input value will be set to `NA`, because they depend on the unknown missing value. In case of `na_rm = FALSE` the `NA` values are treated like zeros.

```
#include <Rcpp.h>
using namespace Rcpp;

// [[Rcpp::export]]
NumericVector cumsumC(NumericVector x, bool na_rm = false) {
  int n = x.size();
  NumericVector out(n);
  LogicalVector is_missing = is_na(x);

  if (!na_rm) {
    out[0] = x[0];
    for (int i = 1; i < n; ++i) {
      if (is_missing[i - 1]) {
        out[i] = NA_REAL;
      } else{
        out[i] = out[i - 1] + x[i];
      }
    }
  }
```

```
  if (na_rm) {
    if (is_missing[0]) {
      out[0] = 0;
    } else {
      out[0] = x[0];
    }
    for (int i = 1; i < n; ++i) {
      if (is_missing[i]) {
        out[i] = out[i-1] + 0;
      } else {
        out[i] = out[i-1] + x[i];
      }
    }
  }

  return out;
}
```

```
cumsumC(c(1, NA, 2, 4))
#> [1]  1 NA NA NA
cumsumC(c(1, NA, 2, 4), na_rm = TRUE)
#> [1] 1 1 3 7
```

The `diffC()` implementation will return an `NA` vector of length `length(x) - lag`, if the input vector contains a missing value. In case of `na_rm = TRUE`, the function will return an `NA` for every difference with at least one `NA` as input.

```
#include <Rcpp.h>
using namespace Rcpp;

// [[Rcpp::export]]
NumericVector diffC(NumericVector x, int lag = 1,
                    bool na_rm = false) {
  int n = x.size();

  if (lag >= n) stop("`lag` must be less than `length(x)`.");

  NumericVector out(n - lag);

  for (int i = lag; i < n; i++) {
    if (NumericVector::is_na(x[i]) ||
        NumericVector::is_na(x[i - lag])) {
```

```
      if (!na_rm) {
        return rep(NumericVector::create(NA_REAL), n - lag);
      }
      out[i - lag] = NA_REAL;
      continue;
    }
    out[i - lag] = x[i] - x[i - lag];
  }

  return out;
}
```

```
diffC(c(1, 3, NA, 10))
#> [1] NA NA NA
diffC(c(1, 3, NA, 10), na_rm = TRUE)
#> [1] 2 NA NA
```

25.5 Standard Template Library

To practice using the STL algorithms and data structures, implement the following using R functions in C++, using the hints provided:

Q1: `median.default()` using `partial_sort`.

A: The median is computed differently for even or odd vectors, which we allow for in the function below.

```
#include <algorithm>
#include <Rcpp.h>
using namespace Rcpp;

// [[Rcpp::export]]
double medianC(NumericVector x) {
  int n = x.size();

  if (n % 2 == 0) {
    std::partial_sort (x.begin(), x.begin() + n / 2 + 1, x.end());
    return (x[n / 2 - 1] + x[n / 2]) / 2;
  } else {
```

```
    std::partial_sort (x.begin(), x.begin() + (n + 1) / 2, x.end());
    return x[(n + 1) / 2 - 1];
  }
}
```

Q2: `%in%` using `unordered_set` and the `find()` or `count()` methods.

A: We use the `find()` method and loop through the `unordered_set` until we find a match or have scanned the entire set.

```
#include <Rcpp.h>
#include <unordered_set>
using namespace Rcpp;

// [[Rcpp::export]]
LogicalVector inC(CharacterVector x, CharacterVector table) {
  std::unordered_set<String> seen;
  seen.insert(table.begin(), table.end());

  int n = x.size();
  LogicalVector out(n);
  for (int i = 0; i < n; ++i) {
    out[i] = seen.find(x[i]) != seen.end();
  }

  return out;
}
```

Q3: `unique()` using an `unordered_set` (challenge: do it in one line!).

A: The `insert()`-method will return if an equivalent element already exists. If a new element is inserted, we will add it to the (unique) return vector of our function.

```
#include <Rcpp.h>
#include <unordered_set>
using namespace Rcpp;

// [[Rcpp::export]]
NumericVector uniqueC(NumericVector x) {
  std::unordered_set<int> seen;
  int n = x.size();

  std::vector<double> out;
```

```
  for (int i = 0; i < n; ++i) {
    if (seen.insert(x[i]).second) out.push_back(x[i]);
  }

  return wrap(out);
}


// As a one-liner
// [[Rcpp::export]]
std::unordered_set<double> uniqueCC(NumericVector x) {
  return std::unordered_set<double>(x.begin(), x.end());
}
```

Q4: `min()` using `std::min()`, or `max()` using `std::max()`.

A: We will implement `min()` by iterating over the vector and recursively comparing each element to the current minimum value.

```
#include <Rcpp.h>
using namespace Rcpp;

// [[Rcpp::export]]
double minC(NumericVector x) {
  int n = x.size();
  double out = x[0];

  for (int i = 0; i < n; i++) {
    out = std::min(out, x[i]);
  }

  return out;
}
```

Q5: `which.min()` using `min_element`, or `which.max()` using `max_element`.

A: To implement `which.min()`, we will first locate the `min_element` and then compute the `distance()` to it (starting from the beginning of the vector).

```
#include <Rcpp.h>
#include <algorithm>
#include <iterator>
using namespace Rcpp;
```

```
// [[Rcpp::export]]
double which_minC(NumericVector x) {
  int out = std::distance(
    x.begin(), std::min_element(x.begin(), x.end())
  );

  return out + 1;
}
```

Q6: `setdiff()`, `union()`, and `intersect()` for integers using sorted ranges and `set_union`, `set_intersection` and `set_difference`.

A: The structure of the three functions will be very similar.

We first sort both input vectors. Then we apply the respective `set_union`, `set_intersection` or `set_difference` function. After that, the result will be between the iterators `tmp.begin()` and `out_end`. To retrieve the result, we loop once through the range between `tmp.begin()` and `out_end` in the last part of each function.

The set operations in base R will discard duplicated values in the arguments. We achieve a similar behaviour by introducing a deduplication step, which omits values that match their predecessor. For the symmetric set functions `unionC` and `intersectC` this step is implemented for the output vector. For `setdiffC` the deduplication is applied to the first input vector.

```
#include <Rcpp.h>
#include <unordered_set>
#include <algorithm>
using namespace Rcpp;

// [[Rcpp::plugins(cpp11)]]
// [[Rcpp::export]]
IntegerVector unionC(IntegerVector x, IntegerVector y) {
  int nx = x.size();
  int ny = y.size();

  IntegerVector tmp(nx + ny);

  std::sort(x.begin(), x.end()); // unique
  std::sort(y.begin(), y.end());

  IntegerVector::iterator out_end = std::set_union(
    x.begin(), x.end(), y.begin(), y.end(), tmp.begin()
  );
```

```
  int prev_value = 0;
  IntegerVector out;
  for (IntegerVector::iterator it = tmp.begin();
       it != out_end; ++it) {
    if ((it != tmp.begin())  && (prev_value == *it)) continue;

    out.push_back(*it);

    prev_value = *it;
  }

  return out;
}

// [[Rcpp::export]]
IntegerVector intersectC(IntegerVector x, IntegerVector y) {
  int nx = x.size();
  int ny = y.size();

  IntegerVector tmp(std::min(nx, ny));

  std::sort(x.begin(), x.end());
  std::sort(y.begin(), y.end());

  IntegerVector::iterator out_end = std::set_intersection(
    x.begin(), x.end(), y.begin(), y.end(), tmp.begin()
  );

  int prev_value = 0;
  IntegerVector out;
  for (IntegerVector::iterator it = tmp.begin();
       it != out_end; ++it) {
    if ((it != tmp.begin()) && (prev_value == *it)) continue;

    out.push_back(*it);

    prev_value = *it;
  }

  return out;
}

// [[Rcpp::export]]
```

```
IntegerVector setdiffC(IntegerVector x, IntegerVector y) {
  int nx = x.size();
  int ny = y.size();

  IntegerVector tmp(nx);

  std::sort(x.begin(), x.end());

  int prev_value = 0;
  IntegerVector x_dedup;
  for (IntegerVector::iterator it = x.begin();
       it != x.end(); ++it) {
    if ((it != x.begin()) && (prev_value == *it)) continue;

    x_dedup.push_back(*it);

    prev_value = *it;
  }

  std::sort(y.begin(), y.end());

  IntegerVector::iterator out_end = std::set_difference(
    x_dedup.begin(), x_dedup.end(), y.begin(), y.end(), tmp.begin()
  );

  IntegerVector out;
  for (IntegerVector::iterator it = tmp.begin();
       it != out_end; ++it) {
    out.push_back(*it);
  }

  return out;
}
```

Let's verify, that these functions work as intended.

```
# input vectors include duplicates
x <- c(1, 2, 3, 3, 3)
y <- c(3, 3, 2, 5)

union(x, y)
#> [1] 1 2 3 5
unionC(x, y)
```

```
#> [1] 1 2 3 5

intersect(x, y)
#> [1] 2 3
intersectC(x, y)
#> [1] 2 3

setdiff(x, y)
#> [1] 1
setdiffC(x, y)
#> [1] 1
```

Bibliography

Stefan Milton Bache and Hadley Wickham. *magrittr: A Forward-Pipe Operator for R*, 2020. URL `http://magrittr.tidyverse.org/`.

Douglas Bates and Dirk Eddelbuettel. Fast and elegant numerical linear algebra using the RcppEigen package. *Journal of Statistical Software*, 52(5): 1–24, 2013. URL `https://www.jstatsoft.org/v52/i05/`.

Rasmus Bååth. The state of naming conventions in r. *The R Journal*, 4(2):74–75, 2012. doi: 10.32614/RJ-2012-018. URL `https://doi.org/10.32614/RJ-2012-018`.

Winston Chang. *R6: Encapsulated Classes with Reference Semantics*, 2020. URL `https://github.com/r-lib/R6`.

Dirk Eddelbuettel. *anytime: Anything to 'POSIXct' or 'Date' Converter*, 2020. URL `https://CRAN.R-project.org/package=anytime`.

Marco Enea. *speedglm: Fitting Linear and Generalized Linear Models to Large Data Sets*, 2017. URL `https://CRAN.R-project.org/package=speedglm`.

Garrett Grolemund and Hadley Wickham. Dates and times made easy with lubridate. *Journal of Statistical Software*, 40(3):1–25, 2011. URL `https://www.jstatsoft.org/v40/i03/`.

Lionel Henry and Hadley Wickham. *purrr: Functional Programming Tools*, 2020a. URL `https://github.com/tidyverse/purrr`.

Lionel Henry and Hadley Wickham. *rlang: Functions for Base Types and Core R and 'Tidyverse' Features*, 2020b. URL `https://github.com/r-lib/rlang`.

Jim Hester. *bench: High Precision Timing of R Expressions*, 2020. URL `https://github.com/r-lib/bench`.

David James and Kurt Hornik. *chron: Chronological Objects which Can Handle Dates and Times*, 2020. URL `https://CRAN.R-project.org/package=chron`. R package version 2.3-56. S original by David James, R port by Kurt Hornik.

Max Kuhn. *modeldb: Fits Models Inside the Database*, 2020. URL `https://CRAN.R-project.org/package=modeldb`.

Thomas Lumley. *biglm: Bounded Memory Linear and Generalized Linear Models*, 2020. URL https://CRAN.R-project.org/package=biglm.

Katharine Mullen, David Ardia, David Gil, Donald Windover, and James Cline. DEoptim: An R package for global optimization by differential evolution. *Journal of Statistical Software*, 40(6):1–26, 2011. URL http://www.jstatsoft.org/v40/i06/.

Kirill Müller and Lorenz Walthert. *styler: Non-Invasive Pretty Printing of R Code*, 2020. URL http://styler.r-lib.org.

John C. Nash. On best practice optimization methods in R. *Journal of Statistical Software*, 60(2):1–14, 2014. URL http://www.jstatsoft.org/v60/i02/.

John C. Nash and Ravi Varadhan. Unifying optimization algorithms to aid software system users: optimx for R. *Journal of Statistical Software*, 43(9): 1–14, 2011. URL http://www.jstatsoft.org/v43/i09/.

Yixuan Qiu, Sreekumar Balan, Matt Beall, Mark Sauder, Naoaki Okazaki, and Thomas Hahn. *RcppNumerical: 'Rcpp' Integration for Numerical Computing Libraries*, 2019. URL https://CRAN.R-project.org/package=RcppNumerical.

R Core Team. *R: A Language and Environment for Statistical Computing*. R Foundation for Statistical Computing, Vienna, Austria, 2020. URL https://www.R-project.org/.

David Robinson, Alex Hayes, and Simon Couch. *broom: Convert Statistical Objects into Tidy Tibbles*, 2020. URL https://github.com/tidymodels/broom.

Simon Urbanek. *fasttime: Fast Utility Function for Time Parsing and Conversion*, 2016. URL https://CRAN.R-project.org/package=fasttime.

Simon Urbanek. *fastmatch: Fast match() function*, 2017. URL https://CRAN.R-project.org/package=fastmatch.

Hadley Wickham. *Advanced R*. Chapman and Hall/CRC, Boca Raton, Florida, first edition, 2014. URL http://adv-r.had.co.nz/.

Hadley Wickham. *ggplot2: Elegant Graphics for Data Analysis*. Springer-Verlag New York, 2016. ISBN 978-3-319-24277-4. URL https://ggplot2.tidyverse.org.

Hadley Wickham. *lobstr: Visualize R Data Structures with Trees*, 2019a. URL https://github.com/r-lib/lobstr.

Hadley Wickham. *sloop: Helpers for 'OOP' in R*, 2019b. URL https://github.com/r-lib/sloop.

Hadley Wickham. *Advanced R*. Chapman and Hall/CRC, Boca Raton, Florida, second edition, 2019c. URL `https://adv-r.hadley.nz/`.

Hadley Wickham and Jim Hester. *readr: Read Rectangular Text Data*, 2020. URL `https://github.com/tidyverse/readr`.

Hadley Wickham, Mara Averick, Jennifer Bryan, Winston Chang, Lucy D'Agostino McGowan, Romain François, Garrett Grolemund, Alex Hayes, Lionel Henry, Jim Hester, Max Kuhn, Thomas Lin Pedersen, Evan Miller, Stephan Milton Bache, Kirill Müller, Jeroen Ooms, David Robinson, Dana Paige Seidel, Vitalie Spinu, Kohske Takahashi, Davis Vaughan, Claus Wilke, Kara Woo, and Hiroaki Yutani. Welcome to the tidyverse. *Journal of Open Source Software*, 4(43):1686, 2019. doi: 10.21105/joss.01686.

Hadley Wickham, Romain François, Lionel Henry, and Kirill Müller. *dplyr: A Grammar of Data Manipulation*, 2020a. URL `https://github.com/tidyverse/dplyr`.

Hadley Wickham, Lionel Henry, and Davis Vaughan. *vctrs: Vector Helpers*, 2020b. URL `https://github.com/r-lib/vctrs`.

Yihui Xie. *bookdown: Authoring Books and Technical Documents with R Markdown*. Chapman and Hall/CRC, Boca Raton, Florida, 2016. URL `https://github.com/rstudio/bookdown`. ISBN 978-1138700109.